THE INTERSTELLAR MEDIUM IN GALAXIES

ASTROPHYSICS AND SPACE SCIENCE LIBRARY

VOLUME 219

THE INTERSTELLAR MEDIUM IN GALAXIES

Edited by

J. M. VAN DER HULST

Kapteyn Astronomical Institute,
Groningen, The Netherlands

KLUWER ACADEMIC PUBLISHERS

DORDRECHT / BOSTON / LONDON

A C.I.P. Catalogue record for this book is available from the Library of Congress.

ISBN 0-7923-4676-9

Published by Kluwer Academic Publishers,
P.O. Box 17, 3300 AA Dordrecht, The Netherlands.

Sold and distributed in the U.S.A. and Canada
by Kluwer Academic Publishers,
101 Philip Drive, Norwell, MA 02061, U.S.A.

In all other countries, sold and distributed
by Kluwer Academic Publishers,
P.O. Box 322, 3300 AH Dordrecht, The Netherlands.

Cover Caption

The distribution of HI in the nearby galaxy M33 shown on the front cover illustrates extremely well the complexity of the morphology of the interstellar medium. The picture covers 1.0 x 1.5 degrees, roughly 1.6 times the optical size of the galaxy. The HI shows a network of spiral arm segments, filaments and shell-like structures. The HI is only one of many tracers of the interstellar medium and the precise morphology one sees depends on which physical state of the interstellar medium one considers. This book covers many states of the interstellar medium, from the coldest to the hottest phases. Though most of the data discussed in this volume has much less detail because the objects under study are much farther away than M33, it is good to keep this complex picture in mind. This HI image has been obtained with the Westerbork Synthesis Radio Telescope and has a linear resolution of 100 parsecs (original data from Deul, E.R. & van der Hulst, J.M., *Astronomy and Astrophysics Supplements,* 67, 509, 1987, reprocessed by Kolkman & van der Hulst).

Printed on acid-free paper

Printed in the Netherlands

Table of Contents

PREFACE

It has been more than five decades ago that Henk van de Hulst predicted the observability of the 21-cm line of neutral hydrogen (H I). Since then use of the 21-cm line has greatly improved our knowledge in many fields and has been used for galactic structure studies, studies of the interstellar medium (ISM) in the Milky Way and other galaxies, studies of the mass distribution of the Milky Way and other galaxies, studies of spiral structure, studies of high velocity gas in the Milky Way and other galaxies, for measuring distances using the Tully-Fisher relation etc.

Regarding studies of the ISM, there have been a number of instrumental developments over the past decade: large CCD's became available on optical telescopes, radio synthesis offered sensitive imaging capabilities, not only in the classical 21-cm H I line but also in the mm-transitions of CO and other molecules, and X-ray imaging capabilities became available to measure the hot component of the ISM. These developments meant that Milky Way was no longer the exclusive playground for ISM studies and that by reaching out to other galaxies astronomy had gained the advantage of having an "outside" view, though at the expense of giving up some linear resolution. Studies of the ISM in other galaxies are intimately connected to studies of the ISM in the Milky Way. First of all because we examine other galaxies with the detailed knowledge of the Galactic ISM in the back of our minds, and secondly because we always try to see whether what we find in other galaxies can help to interpret the structures and features of the Galactic ISM which we do not yet understand.

Significant progress has been made in the past decade and it appeared imminent at this time to try to synthesise the current knowledge in the area of studies of the different components and aspects of the ISM in nearby galaxies. This notion led to the concept of this volume on the "Interstellar Medium in Galaxies". Seven experts in the field provide an excellent overview of the current knowledge and future progress in the area. The chapters are roughly ordered by the temperature and density of the respective components: the dust, the cold molecular gas, the cool gas probed by the 21-cm H I transition, the warm ionized component and finally the hot component which is traced by its X-ray emission. The last two chapters cover the composition of the ISM and the relation between properties of the ISM and the star formation process. This volume is meant both for experts in the field as background material and for others as introductory reading to get familiar with the subject.

I would like to thank the authors of the chapters for providing me with excellent material and for bearing with me when trying to get editing of this volume out of the way. When we originally discussed the possibility of putting together a book like this we were very ambitious. Not all the original ideas were realized. Yet I think that thanks to the authors' contributions this volume will be useful for a large fraction of the entire astronomy community.

Thijs van der Hulst
Groningen, May 15, 1997

DUST IN THE DISKS OF GALAXIES

M. SAUVAGE
Service d'Astrophysique, DAPNIA/SAp,
C.E. Saclay, C.E.A.
91191 Gif-sur-Yvette CEDEX,
France

1. Introduction

As astronomers involved in studies of interstellar dust in various environments are about to head into ISO's decade, it seems to me both a good and a risky idea to try and summarize our current knowledge of the properties of dust in galactic disks. Indeed it is always interesting to sort out our various findings to see which questions were answered, which still need research and which have arisen. It is however never safe to make a bet on the future, as our expectations are often based on a description of reality strongly biased by the limitations of available instruments. This is even more unsafe a few months before the first light of a "ground-breaking" telescope.

I will nevertheless perform such an exercise, hoping no to be proven too wrong or too short-sighted during the next years.

It has now been more than ten years since IRAS, by its sky-survey, opened wide the field of extragalactic dust by revealing that most galaxies harbor large quantities of dust radiating in the mid to far infrared. Since then, this field has developed dramatically, not because dust is an important fraction of the interstellar medium (it represents only $\approx$1% of the gas mass of a spiral galaxy) but because:

- contrary to other tracers of the interstellar medium (ISM) it radiates continuously over $\approx$3 decades of wavelength, making it accessible to a growing number of instruments,
- it participates in one of the most important processes of galaxy evolution: star formation.

IRAS and ground-based telescopes (telescopes working in the 10-20 μm regime, KAO, and submillimeter antennae) have now revealed the basic

J. M. van der Hulst (ed.), The Interstellar Medium in Galaxies, 1–31.

properties of dust in galaxies, and with the help of a new generation of instruments, we will try to go into the details of this phase. Among these basic properties, three are of particular importance.

We know that dust is widespread in galaxies. Apart from elliptical galaxies where it does not appear to be extended (this may only be due to our instrumental sensitivity), the dust phase generally extends up to at least half the optical radius (taken at the 25th mag.$''^{-2}$) of spiral galaxies. Consequently it can probably not be considered as a specific phase of the ISM in the same respect as H_2 or H I , as it is probably mixed both with molecular and atomic gas. As I will discuss later, dust is present in various states in a galaxy which can be characterized by their excitation properties.

We also know that dust appears quite early in the "life" of a galaxy: IRAS F10214+4724, the most luminous galaxy known yet (Rowan-Robinson *et al.* [1991]), has been discovered through its infrared emission and is now known to harbor large quantities of dust. Considering that it has a redshift of 2.286, one sees that large amounts of dust are not restricted to the local Universe.

Finally it is also clear that dust either plays an important part in, or at least is strongly linked to, the star formation process. Star formation is, in normal spiral galaxies, the most important provider of UV radiation. Dust in turn is extremely sensitive to UV photons and therefore can act as an efficient tracer of star formation. In the following, however, I will show that the link between dust emission and star formation still requires some clarification.

Aside from these general trends, there are a number of details that are still missing and that hamper a clear understanding of dust emission. Among others, let me cite the following:

- The detailed chemical composition of dust components is only vaguely known. As a result, there are large uncertainties in the spectral properties used to derived physically meaningful quantities from dust emission.
- The spatial distribution of dust inside galaxies is not well known. So far our knowledge is limited by the sensitivity of our instruments, and very few information on the host galaxy can be drawn from the comparison of gas and dust spatial distribution.
- Dust is present in a galaxy in quite a number of physical states, as is evident from studies of our own Galaxy. In external galaxies, however, it is quite difficult to accurately describe and map these different phases due to the rather low spectral and spatial resolution of our instruments. Thus the physics of the dusty ISM if for a large part unknown.

Let me point out that this situation is not unlike that encountered when studying the molecular phase of the ISM through CO emission, as there again, the physics relating the tracer to the ISM is still mainly to be built.

In light of these points, I have organized this review in the three following parts: first, as most of the uncertainties arise from our limited knowledge of the constituents of dust, I will discuss the proposed dust models, with an emphasis on the consequences each of them has on the various derived quantities (mass, temperature, ...). Then I will try to identify the major sources of heating as well as the different phases of dust that can be isolated in a galaxy. I will try to relate each of these phases to already known phases of the ISM. Finally, in a more speculative part of this paper, I will discuss the issue of formation and processing of dust grains in galaxies, as well as try to outline what the very near future may bring.

2. Dust models and their consequences

As this is clearly an important topic, a number of reviews are entirely devoted to discussing the various properties and capabilities of proposed dust models. Among these I refer the reader to Mathis (1990) who gives a detailed analysis of what extinction effects tell us about the dust, and to Puget & Léger (1989) who present the main properties of dust models containing transiently heated grains. Also of very high interest on this topic are the conference proceedings *Molecules and grains in space* ([1994], I. Nenner ed., AIP Press:New York).

To assess the role of dust in the ISM, one needs the spatial distribution of dust temperature, T_d, as well as that of the dust mass, M_d. To be even more precise, one would like to have the distribution of dust mass according to both space and dust temperature, so that its physical state can be characterized (it is likely that there is a range of T_d on any given line of sight inside a galaxy).

Yet up to now, most of our measurements consist of broad band photometry of dust emission, be it from IRAS or ground-based telescopes. We are therefore requested to use dust models to establish a connection between what we measure and what we wish to derive. Although some agreement has been reached in the past few years on which model(s) should be used, some consequences of the choice made are not always clear to the users and can produce spurious results. Furthermore, the data have shown the need for some refinement in the models, on which the debate is still ongoing.

2.1. A COMMON FEATURE OF DUST MODELS: LARGE GRAINS

Before the infrared era, largely created by IRAS, dust was mainly known by its absorption effect on the light of nearby stars. This effect it synthetically characterized by the well known extinction curve, Figure 1, where the

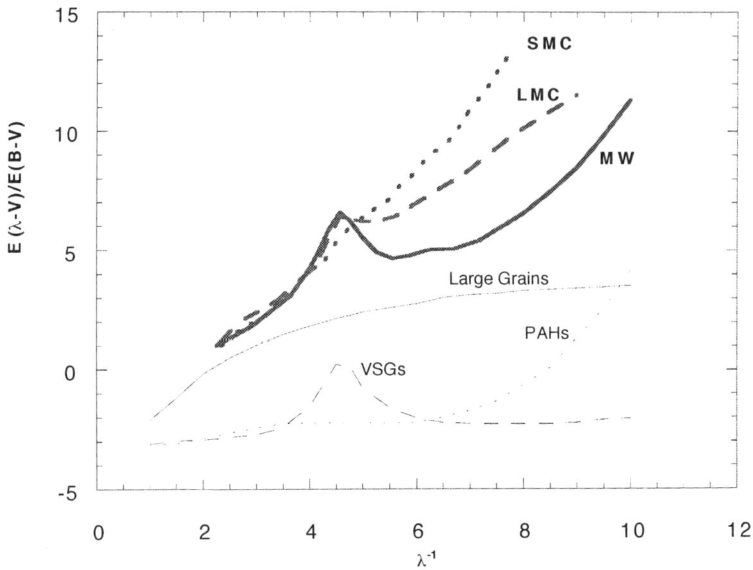

Figure 1. The standard extinction curve, an average of extinction curves along Galactic lines of sight (from Savage & Mathis [1979]). Also plotted in dotted lines are the only two other well-known galactic extinction curves, that of the Large and Small Magellanic Clouds (respectively from Koornneef & Code [1981] and Prévot *et al.* [1984]). Possible origins for the differences between these three curves are discussed in the text. On this figure is also shown the decomposition in three dust components made by Désert *et al.* (1990) in their attempt to synthesize infrared emission of various objects of the ISM.

normalized extinction, $E(\lambda - V)/E(B - V)$, is plotted against decreasing wavelength.

This curve is generally divided in three parts, a linear rise in the near infrared (NIR) to visible range, a "bump" at 2175 Å ($4.6\,\mu m^{-1}$), and a non-linear UV rise. These three features can vary relative to each other as is evident from the comparison of the Galactic curve to that of the LMC and the SMC: The bump slowly disappear while the UV rise becomes both much more important and linear. This behavior is obviously related to changes in dust composition though it is still unclear which should be invoked (see e.g. Jones *et al.* [1987], and references therein, and Mathis & Cardelli [1992]).

The linear part of the curve, $\propto \lambda^{-1}$, is characteristic of grains of size a comparable to the wavelength, i.e. $a \lesssim 1\,\mu m$. These are the so-called "large grains" of most current dust models. In the first few dust models, e.g. Draine (1981), these grains were made of silicates and graphite. The reason for this is quite straightforward: silicates have unambiguous absorption signatures at $9.7\,\mu m$ and $18\,\mu m$, detected in the mid-70's in a wide variety of environments (see discussions in Savage & Mathis [1979], and Whittet, [1988]). Furthermore, Silicon emission is evident and abundant in evolved stellar spectra making it a likely component of dust. For graphite, the reason is similar: carbon is abundant in the ejecta of evolved stars and graphite has an emission band in the 2200 Å region.

Though it is clear, from the absorption/emission signatures in the NIR, that silicates are present in the interstellar medium, it is now unlikely that pure graphite is indeed a component of interstellar dust. First, condensation of carbon grains in the envelope of an evolved star leads more to amorphous carbon than to graphite, though exposure to UV radiation and particles through the life of the grain may gradually create or enlarge graphitic cores. Second, though graphite has an emission band at 2000 Å, laboratory measurements have been unable to reproduce one of the main properties of the interstellar bump: its full width at half maximum is quite variable (from $0.80\,\mu m^{-1}$ to $1.2\,\mu m^{-1}$), while the central wavelength remains constant (2174±9 Å, Mathis, 1994). Current interstellar dust models still attribute the 2175 Å bump to carbonaceous material but it is either coated, with PAHs or refractory elements such as Fe or Mg, or coal-like such as in Papoular *et al.* (1993), or undefined such as in the PAH models (Désert *et al.* [1990], or Siebenmorgen & Krügel [1992], see later).

Although these grain models can be combined to reproduce the interstellar extinction curve (see Draine & Lee [1984]), a rapid computation shows that they cannot reproduce the whole infrared emission spectrum as observed by IRAS. The size of these grains, typically a few tens of Å to less than $1\,\mu m$, implies that they will reach thermal equilibrium in the interstellar radiation field. Draine & Lee (1984) have shown that, exposed to the mean interstellar radiation field of Mathis *et al.* (1983), silicate and graphite grains of $0.1\,\mu m$ reach temperatures of 19 and 15 K respectively. Though this temperature is compatible with that of cirrus clouds, as deduced from the 60 and 100 μm fluxes, it is too low to explain the continuum emission observed by IRAS at 12 and 25 μm in the same clouds.

As a result, current models do incorporate large silicate and carbon grains, but for a limited part of the spectrum. For models including PAHs (see later), large grains are responsible for the visible linear rise of the extinction curve as well as for a constant fraction of the extinction on the rest of the spectrum. This is shown on Figure 1 by the curve labeled "large grains" reproduced from the model of Désert *et al.* (1990). Given the temperature that these grains reach in most parts of the ISM (10 to 30 K, see section 3), their emission will be prominent mostly in the submillimeter and far infrared (FIR) domains.

This is the reasoning that leads to the use of the IRAS $S(60\ \mu m)/S(100\ \mu m)$ color as an estimator of the dust temperature. Section 2.4 will show how reliable this estimator is.

The geometrical structure as well as exact chemical composition of these large grains is yet unclear. It is unlikely that they are both chemically pure and spherical in geometry, as is assumed generally to derive their spectral properties. On the contrary, there is evidence that their shape may be quite fluffy (see e.g. Mathis & Whiffen [1989]). As for their chemical composition, it is also likely that they contain impurities as they probably grow by accretion of smaller dust grains or molecules. Furthermore, as grains are the likely reaction site for the formation of many molecules, they could be covered by dirty ice mantles (see e.g. Duley & Williams [1994]).

An alternative to models including PAHs or other very small grains has recently been developed by R. Papoular and his team. As this model differs quite strongly from the preceding class in its principle, I will discuss it in a dedicated part of this section. Let me however mention that in this model, all the infrared emission and the absorption is accounted for by a mixture of large silicates and coal-like grains.

2.2. TRANSIENTLY HEATED GRAINS AND UNIDENTIFIED MIR BANDS

In 1985, K. Sellgren published the NIR spectrum of three reflection nebulae, which showed the presence of strong emission bands at 3.3 and 3.4 μm, on top of a continuum of color temperature of $T_c \approx 1000$ K, independent of the distance from the star. This discovery, and that of the ubiquitous 12 μm emission observed by IRAS, were the starting point for a new class of interstellar dust: transiently heated grains. It is now recognized that the 3.3 and 3.4 μm bands are part of a larger family, at the time nicknamed "unidentified infrared bands" (UIB) but which should be now named "over-identified" given the number of proposed carriers. Figure 2 shows NIR to MIR spectra of various objects, from reflection nebulae to H II regions to galaxies that all show the UIB.

The problem raised by Sellgren and IRAS observations is the following: how are we to explain emission that, if produced by a classical grain in thermal equilibrium, implies a temperature incompatible with the heating field available? Furthermore, the temperature that would be derived does not depend on the intensity of the heating field!

To solve the problem raised by the ubiquitous MIR emission, Léger & Puget (1984) proposed that the emission is associated with extremely small grains, or very large molecules (≈ 50 atoms), the polycyclic aromatic hydrocarbons (PAHs). These grains present a number of features that makes them extremely attractive:

- Upon absorption of a single UV photon, they can redistribute the absorbed energy in internal levels that decay through visible and infrared fluorescence. As this process is extremely fast, $\lesssim 10^{-7}$ s, it usually occurs before the next absorption takes place. As a result, the color

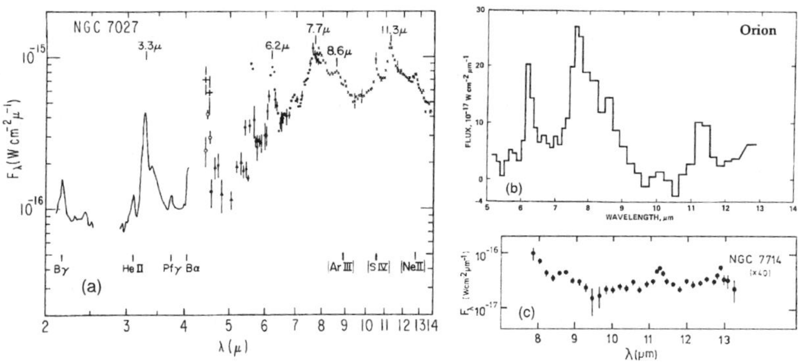

Figure 2. The U in UIB (Unidentified Insterstellar Bands) could also stand for Ubiquitous, as revealed by these three figures, where they appear in three quite different astrophysical sites. (a) The spectra published by Russel (1977) on the reflection nebula NGC 7027. (b) A line of sight in the Orion HII region (Bregman *et al.* [1989]). (c) The interacting galaxy NGC 7714, as observed by Phillips *et al.* (1984). Though the shape and relative intensities of the bands vary from one object to another, the central wavelength are always 3.3, 6.2, 7.7, 8.6, and 11.3 μm. These bands are attributed to C-H and C-C bonds.

temperature of such an emission depends much more on the spectrum of the incident radiation than on its intensity, readily explaining the color-constancy observed by Sellgren.

- They are strong absorbers in the UV and could thus be responsible for the UV rise observed in Figure 1 (the curve labeled PAH corresponds to the phenomenological decomposition of Désert *et al.* [1990] not to the measured absorption curve of PAHs).
- As they contain the functional groups C-C and C-H, they present well defined emission bands corresponding quite nicely to the so-called "unidentified" infrared bands at 3.3, 6.2, 7.7, 8.6 and 11.3 μm.

These three properties alone explain the attention that PAHs have received during the last decade. Indeed, combined with the large grains mentioned earlier, they allow models to nicely fit the whole infrared spectra of a wide variety of objects such as those displayed in Figure 2. Siebenmorgen (1994) present quite a remarkable fit of the M82 spectrum from 2 μm to 3 mm with a model that incorporates PAHs. Note that PAHs are mixed with large grains, and that the whole IR spectrum, from 3 μm to 800 μm can be reproduced with a single dust mixture and a single radiation field. This is not the case in the coal model where the NIR-MIR emission comes from dust located closer to the heating sources than the long-wavelength emitting dust (see section 2.3).

PAHs, however, are not enough to fully reproduce the observed infrared spectrum of interstellar objects.

Their first short-coming is that models incorporating only PAHs and large grains cannot reproduce emission observed in the 25 μm band of IRAS (Sauvage *et al.* [1990]). Indeed PAHs do not emit much in this band while large grains are generally too cold for the peak of their black-body emission to extend to this wavelength. Yet the 25 μm emission of galaxies, for instance, is generally larger than their 12 μm emission. To compensate for that, PAH models are forced to call on to another component, loosely described as Very Small Grains (Puget & Léger [1989]). Their nature is unknown but probably carbonaceous, since, in the models of Désert *et al.* (1990) and Siebenmorgen & Krügel (1992), they are responsible for the 2175 Å bump in the extinction curve (see curve labeled VSG in Figure 1).

Though their nature is unconstrained, VSGs are extremely important as far as derived properties of the dust are concerned: indeed their emission is not only present in the 25 μm band of IRAS, it also makes up most of the 60 μm band emission (see Figure 4 of Désert *et al.* [1990]). As a result the $S(60\ \mu\mathrm{m})/S(100\ \mu\mathrm{m})$ ratio is no longer a reliable measure of the large grains temperature. The mass derived from the same ratio is also no longer a correct measure of the dust mass. These effects will be detailed later (section 2.4).

The second short-coming of PAHs is their absence of continuum emission as measured in the laboratory. As can be seen from Figure 2 the bands themselves make up only a small fraction of the emission in the MIR (a few % typically for the 12 μm band). To correct for this an ad-hoc continuum is added to the Désert *et al.* (1990) model, while Siebenmorgen & Krügel (1992) postulate the existence of PAH clusters to whom they attribute most of the continuum emission.

Third, as individual PAH species have a UV absorption spectra presenting a lot of structure, contrary to the smooth non-linear rise observed in Figure 2, it is still to be demonstrated that a mixture of various PAH species can be responsible for the far UV extinction, as postulated by tenants of the PAH model. As a favorable point on this subject, we mention the work of Lee & Wdowiak (1993) who show that ionized PAHs have much less structure in their UV absorption spectrum than neutral ones (see also next point).

Finally doubts can be raised on the very existence of PAHs in the interstellar medium, as Schlemmer *et al.* (1994) have shown that in the gas phase, small neutral PAHs have an emission spectrum that is markedly different from that shown in Figure 2. It is however quite likely that PAHs would be ionized in the interstellar medium, as their first ionization potential is low (Mathis [1990]). Furthermore, as PAHs have been found in meteorites, they do exist in the ISM, the question being in what amounts.

This shows that, as attractive and powerful as it may by, the PAH model still leaves room for alternate dust models.

Before moving to the next section, let me mention briefly that a number of models resembling in essence the PAH model have been proposed. I will not discuss them here, however, because they have not been developed up to the point where they can actually be used to derive properties of the dust from the observed spectrum, such as the PAH model. I refer the interested reader to the work of e.g. Duley & Williams (1988) on hydrogenated amorphous carbons (HAC) or Sakata *et al.* (1987) on quenched carbonaceous composites (QCC).

2.3. AN ALTERNATIVE: THE COAL MODEL

The coal model is part of a comprehensive attempt at understanding dust in galaxies, from the formation process, to the evolution in the ISM, and of course to the production of observed emission/absorption features. With the PAH/VSG model, it is also one of the few models complete enough to make definite predictions that can be tested with observations.

The original ingredient of the coal model is obviously coal. Classical "astronomical" silicates (Draine & Lee, 1984) are also added to provide the absorption features and the far infrared (FIR, longward of 60 μm) emission. Coal is in fact a generic term describing a whole class of materials where one always find the following structure: a loose assembly of small bricks made of layers of aromatic carbon rings (10 to 20 Å in size, the so-called "basic structural units", hereafter BSUs) on which functional groups are attached. These groups are generally CH, CH_2, COH, or oxygen bridges between BSUs. Coal species are ordered by rank, a property that reflects their age, or evolutionary status. A higher rank implies an increase in the graphitization of the material and in the carbon fraction as compared to O and H. Abundances in typical coals for C, H and O are respectively proportional to 1:0.5:0.02 (Guillois *et al.* [1994a]). Figure 3a shows a typical arrangement of a coal grain. Note that even a coal grain of high rank is not identical to graphite as BSUs are still stochastically arranged with respect to one another.

Given the functional groups present in coals, it comes as no surprise that these grains should produce the UIB. They are attributed to the same transitions as in PAHs. A number of additional bands appear in experimental emission measures on coal but Papoular *et al.* (1991) have shown that these can all be related to the low rank of the coal used. As the rank increases, the spectrum gradually tends toward that observed in the ISM. Figure 3b shows the observed spectrum of one such high ranked coal compared to that of the Red Rectangle.

A further interest of the coal model is its ability to reproduce the

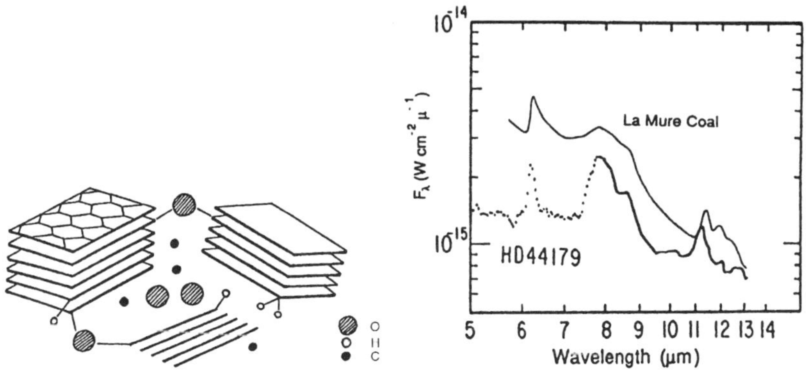

Figure 3. (a) Typical structure of a coal dust grain: small (10 to 20 Å) bricks made of layers of aromatic carbon cycles on which functional groups such as CH, CH_2 or COH are attached. These bricks are loosely packed together, and occasionally an oxygen atom bridges the gap between two bricks. As coal evolves with time, the rank increases, H and O abundances are reduced while the number density of graphitic bricks increases. Note however that the disordered arrangement of these bricks is maintained, and the material does not turn into a graphite grain (figure adapted from Guillois *et al.* [1994a]). (b) The expected spectrum of a high rank coal sample (from the La Mure seam) compared to the emission of the Red Rectangle (HD 44179). Note that due to the experimental method used, the slope of the underlying continuum is not correctly estimated. One can see that the fit is quite reasonably good, considering that it involves no adjustment of free parameter (figure taken from Guillois *et al.* [1994b]).

extinction curve. Since, as I mentioned before, models involving PAHs and VSG have to perform an ad-hoc fit of this curve, this represents a significant improvement. Once again, as the material contains graphite "islands" it is not surprising to find that the 2175 Å bump can be produced by coal grains. On this point, it is of further interest to note that the position of the bump, as well as its intensity are also a function of the coal's rank: the higher the rank, the stronger and closer to the observed wavelength the bump (Papoular *et al.* [1993]). Consequently, evolved coals can account for the bump while the continuum visible extinction is produced by less evolved grains. This situation, where different grain species coexist is more compatible with the current picture of the dust phase, where grains are impacted by charged particles, atoms and molecules and thus are very likely to undergo structural changes in the ISM.

As for the non-linear UV rise of the extinction curve, Papoular *et al.* (1993) have shown that it can be accounted for by the red wing of the only other absorption band of coal in the far UV range.

Concerning the thermal properties of coal, they can all be traced back to the fact that it is a semi-conductor. As a result, the absorptivity dramatically increases from the near infrared to the visible, contrary to the slower λ^{-1}

increase observed for silicate grains. The equilibrium temperature reached by a coal grain in a given radiation field is thus much higher than that reached by an equivalent silicate grain. For instance, Guillois *et al.* (1994b) have shown that coal grains, of size 200 to 2000 Å, in the radiation field of NGC 7027, would reach an equilibrium temperature of $\approx$ 400 K while normal grains would only reach 200 K. Furthermore, for this value of the equilibrium temperature, the color temperature between 2.2 and 4.8 μm is $\approx$1000 K, just as observed by Sellgren (1985) on such nebulae. Puget *et al.* (1995) have shown however that the illuminating stars of the well known nebulae NGC 7023 and NGC 2023 do not provide enough energy to heat coal grains at the required temperature. Although this fact is interpreted by Puget *et al.* as invalidating the coal model, it could mean that either the size distribution of coal grains extends down to sizes where transient heating is important, or embedded stars also contribute to the heating, as is frequently the case in such objects.

This model has however an important shortcoming concerning the 3.3 μm emission band. For coal to emit substantial amounts of radiation in this band, high temperatures (i.e. $\gtrsim$ 300 K) are required. These temperatures can be reached in reflection nebulae or H II regions but not in the general interstellar medium. If it is clearly shown that diffuse 3.3 μm emission is detected in the ISM, transient heating of smaller coal grains/PAHs would have to be included. However, it is yet too early to conclude either from IRAS measurements or from the balloon-borne AROME experiment (Giard *et al.* [1989]), that the band emission comes indeed from the diffuse interstellar medium. The coarse spatial and spectral resolution of these instruments is compatible with the emission originating from hot nebulae. A fraction of only $\approx 10^{-4,-5}$ $M_{\rm d}$ close to these nebulae is enough to account for the detected emission (Papoular et al., 1995). The validity of this assumption will be tested by ISO.

2.4. UNCERTAINTIES AND MEASUREMENTS

From the previous sections it is quite clear where the main uncertainties in interpreting observations lie: as no agreement on the dust composition has been reached, and as none of the models possess accurately known spectral properties, it remains difficult to deduce from infrared and millimetric fluxes the bulk properties of the emitting dust.

Of special interest is to determine the dust mass in an object, or derive the approximate temperature(s) of the dust. In galaxies, a dominant fraction of the dust mass is rather cold. As a result, dust masses are determined from FIR and submillimetric fluxes. At these wavelengths the emission of all dust models is a that of a grey body, which eases the determination of $M_{\rm d}$.

However, the exponent of the emissivity law, $\kappa_\nu \propto \nu^{-n}$, in this wavelength range is not well constrained. n is usually taken between 1 and 2 but even this narrow range has large consequences on M_d or T_d (a factor of 15 on M_d, see Kwan & Xie [1992]). Most of the controversy on the existence of a cold (10-15 K) dust phase in galaxies can be traced back to this uncertainty in the exponent of κ_ν. This is comprehensively studied by Kwan & Xie (1992).

Possible constraints on n can come either from laboratory measurements, which are complex and imply the choice of a dust model, or from observations: obtaining FIR/Submm measurements on simple objects where the dust temperature distribution is not expected to cover a wide range (i.e. not galaxies...) or studying the variation of the FIR emission properties as the elemental abundances changes (e.g. Sauvage *et al.* 1990).

Further uncertainties are introduced by the instruments we use to study dust, and mainly by IRAS filters. Though these are quite straightforward to estimate, they have been surprisingly overlooked. One frequently finds in the literature a sentence quoting the unexpected difference between the equilibrium temperature of dust grains in the interstellar radiation field (ISRF, $\approx$ 20 K), and that observed by IRAS on quiescent dust clouds ($\approx$ 25 − 30 K). The IRAS temperature is generally computed with (see e.g. Thuan & Sauvage, 1992):

$$T_\mathrm{d} = \frac{95.94\,\mathrm{K}}{\ln\left[(1.67)^{3+n}\frac{S(100\,\mu\mathrm{m})}{S(60\,\mu\mathrm{m})}\right]} \tag{1}$$

This difference is interpreted in various ways that I will discuss later. Of importance here is the major flaw that this reasoning contains: the two temperatures should only be equal if (1) the instrument is "perfect" (in the case of IRAS, perfection would be monochromatic bands), and (2) the fluxes used actually come from only one component in thermal equilibrium. None of these point is true. The IRAS bands are far from monochromatic (they are in fact a few tens of μm large). I also have shown previously that whatever the model used, only the 100 μm band can be considered as coming from grains in thermal equilibrium (though see section 3), while all other bands, the 60 μm band in particular, collect emission from more than one species, emission that could be due to transient heating. As an example, in the model of Désert *et al.* (1990) for dust in the ISRF, the equilibrium temperature of the large dust grains is 21 K while the temperature deduced from the predicted fluxes through (1) is 25 K. This clearly shows that T_d is at best only an indicator of the dust temperature (see section 3).

With that sort of uncertainties on T_d, what is the validity of M_d derived from the IRAS 60 μm and 100 μm fluxes?

From the relatively high temperature deduced from IRAS and from the observation that hot dust, even in small amounts, rapidly dominates the emission, some authors have argued that IRAS gives a relatively good measure of the amount of hot dust in galaxies. This is however not the case if, as I will show later, cold dust exists in large amounts in galaxies. For instance, Draine (1990) takes the example of a mixture of 0.4% of dust at 36 K, a temperature likely to be observed around H II regions, and 99.6% of cold dust at 18 K. If one simply assume that the resulting 60 and 100 μm fluxes come from only one dust phase, one will measure a dust temperature T_d of 23-24 K, i.e. different from both true temperatures, and a dust mass M_d of only 20% of the total dust mass, i.e. 50 times the mass of the hot component and 1/5 that of the cold component. Note that this distinction into two phases should only be considered for its demonstrative value, while in reality it is quite likely that there is a continuous distribution of dust temperatures. If transiently heated grains contribute to the 60 μm emission, the computed dust mass will be even further from reality.

Thus M_d is also at best an indication of the total amount of dust present in an object.

The above consideration illustrates the limitations imposed by the IRAS survey. It is clear that spectra in the FIR domain provided by ISO will allow us to better determine such basic properties of the dust such as its mass and temperature(s). This will surely improve our knowledge of the role of dust in galaxies.

3. Heating sources and phases of the dust

Although the infrared domain offers us a direct view of dust in galaxies, it is now realized that dust is not an isolated phase of the ISM, but is rather mixed with the cold atomic gas, the molecular gas or the hot neutral gas in photodissociation regions (PDRs). Since dust is present in these various environments, a question appears: what are the sources of energy that heat the dust we observe? Behind this question is the more important one: can dust emission be used to trace something else than dust itself, i.e. star formation, the interstellar medium, the type/history of the host galaxy, ...?

3.1. STAR FORMATION AS A HEAT SOURCE FOR THE DUST PHASE?

Star formation is an obvious source of energy for dust emission. This is amply demonstrated by the fact that starburst and interacting galaxies are extremely strong FIR emitters, that the Orion nebula is one of the strongest IR source in the sky, that massive O stars produce high levels of high energy radiation, and that dust is very sensitive to UV radiation (but not only, see section 2.3).

Based on these observations, it would be tempting to assume that star formation is always the dominant only source of energy for dust emission, so that $L_{\rm FIR}$ is a direct tracer of the star formation rate.

This is however not consistent with another set of observations that indicate that older stars also contribute to dust heating. Let me cite the following:

- Dust emitting at FIR wavelength, i.e. mostly large grains, is more sensitive to the integrated intensity of the ISRF, not to its spectral shape (see Figure 1).
- In the Milky Way, even taking into account uncertainties in star counts, there are not enough O stars to explain the observed FIR emission (Cox & Mezger [1988]). This is true even if all their UV flux is reprocessed by dust, a situation which does not occur, the local reprocessing factor being $\approx$ 20% (Leisawitz & Hauser [1988]).
- The ISRF in the solar neighborhood, as derived by Mathis *et al.* (1983) comprises only a very small contribution from young massive stars. The dominant contribution is that of older red stars. Therefore emission of dust in such a quiescent environment will not be very sensitive to the young stellar population of the host galaxy.
- The IR colors of normal galaxies are observed to span a continuous range from those characteristic of H II regions to those of neutral H I clouds (e.g. Helou [1986]); this behavior can only be understood if, in some galaxies, older stars contribute the main part of the heating flux (Sauvage & Thuan [1994]).

These general trends show that older stars cannot always be neglected in dust heating. More recently, detailed studies of nearby galaxies have confirmed this trend. They reveal that the FIR spectrum is markedly different between regions of star formation and quiescent regions: the dust spectrum is hotter in star forming regions, see Figure 4.

Also worth mentioning is the arm-interarm contrast in M 51 which decreases from Hα, to FIR, to Blue requiring that another source than high mass stars heats the dust (Rand *et al.* [1992]). Contribution to the heating by older non-ionizing stars is also required to explain the non-linearity of the FIR-radio correlation (e.g. Condon *et al.* [1991], Völk & Xu [1994]).

A KAO study on the early-type spirals NGC 4736 and NGC 3627 by Smith et al. (1994) similarly showed that their FIR emission cannot be powered by the very few young stars they harbor.

From the above mentioned evidences, it is then clear that older stars contribute some heating but it can be argued that, given the difference in

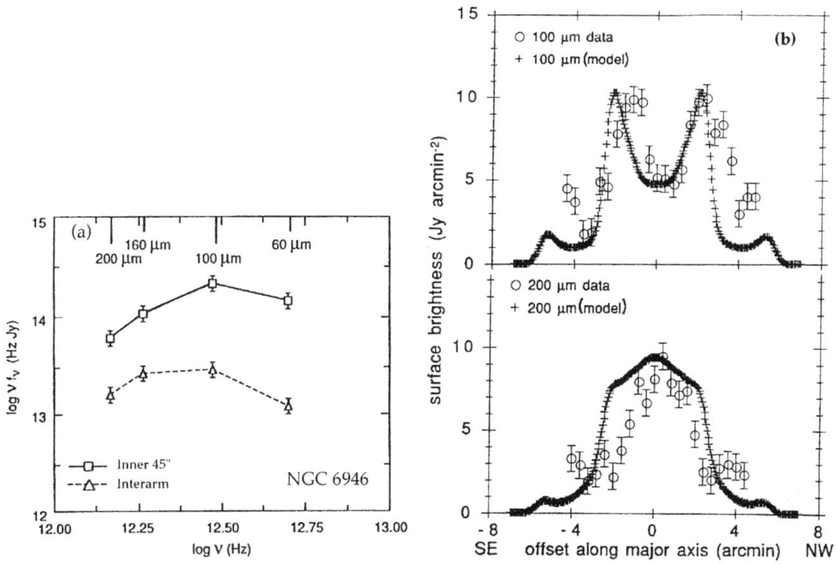

Figure 4. Examples of different FIR spectra observed within one galaxy: (a) In the galaxy NGC 6946, the hotter spectrum corresponds to regions with high levels of star formation (the 45″ inner disk), while the cooler spectrum is that of the interarm regions, composed mainly of old stars (Engargiola [1991]). (b) The emission of the edge-on galaxy NGC 4565 is remarkably well reproduced by a cool 20 K disk and warm 30 K arms (Engargiola & Harper [1992]).

luminosities, high mass stars generally contribute most of it. This is the position supported by some authors, e.g. Devereux & Young ([1991], and subsequent papers). These authors show that the Hα/FIR ratio observed in galaxies (either in samples or in individual galaxies) is that expected from dust illuminated by an 05-B9 star. Taken at face value, this is a very strong argument in favor of a star formation powered FIR emission in most galaxies. There are however two weak points in this comparison: first to derive the expected Hα/FIR ratio from H II regions powered by stars of given type, Devereux & Young assumes that the full bolometric luminosity of the star is converted into FIR radiation while observations and models shows that this conversion factor is only 20% (e.g. Leisawitz & Hauser [1988]). Second, the authors do not discuss the non-linearity observed in the Hα/FIR correlation. Sauvage & Thuan (1992) have shown that this non linearity is mainly due to a varying contribution of cold dust powered by non ionizing stars to the FIR radiation. This varying contribution is related to the morphological type of the considered galaxies (see Figure 5). This interpretation is confirmed by the KAO observations by Smith *et al.* (1994, 1995) on early-type spirals.

Thus the conclusion from this discussion is that, though star formation

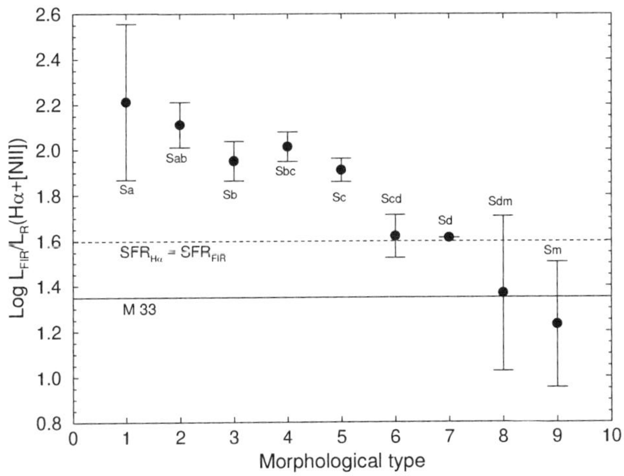

Figure 5. Variation of the $L_{\mathrm{FIR}}/L_{\mathrm{H}\alpha}$ ratio in galaxies of different types. The ratio decreases from early-type to late-type spirals, to reach values observed in M 33 HII regions (Wilson *et al.* [1991]). Also indicated is the ratio where star formation rates derived from FIR and Hα luminosities are equal. The error bars show the error in the mean. The decreasing trend and the agreement of the last data point with the values measured on individual star-forming regions of M33 imply that only in late-type galaxies is the FIR powered by the hot stars giving rise to the Hα flux. In earlier galaxies there is an excess of FIR emission compared to what can be produced by the hot stars. This heating source for this excess has to be the older and much mumerous stars (figure adapted from Sauvage & Thuan [1992]).

is an important source of heating for the dust, it cannot be considered as the only one in normal galaxies. In the latter, older stars, which are much more numerous, can be the most important contributors far from star-forming regions. L_{FIR} is therefore not a reliable tracer of star formation, unless the ratio $L_{\mathrm{FIR}}/L_{\mathrm{H}\alpha}$ is close to the value observed for late-type spiral galaxies.

3.2. PHASES OF DUST IN GALAXIES, A WIDE RANGE OF TEMPERATURES

The short review presented here of arguments for different sources of heating also show that dust is present in various environments with a range of temperatures. In the following I present a description of this range and try to associate each phase with a different component of the ISM. Each of these phases will see its relative importance vary with the type of the galaxy, its stellar population, gas content and star formation activity. This readily explains while some authors (e.g. Eales *et al.* [1989]) rejected the idea that multiple dust phases existed in galaxies, on the basis of samples biased toward e.g. FIR bright galaxies.

In the following subsections I have placed limits to the temperature of each considered phase. These are at best indicative since the only quantity on which it is easy to place limits is the emission wavelength range. Thus the temperature values corresponds to those of black-bodies that would emit in the same wavelength ranges. To give some physical meaning to them, they are similar to that which a large grain such as an "astronomical" silicate grain could reach.

3.2.1. *A hot dust phase - $T_{\rm d} > 50\,K$*

Given its temperature, this phase is more clearly seen in the mid infrared, (from 10 to 40 μm). As the true nature of the dust is yet unresolved (see section 2) the emission collected in this wavelength regime is likely to be a mixture of emission from small particles such as PAHs and emission from hot grains in thermal equilibrium in the vicinity of powerful radiation sources (H II regions, hot stars or compact sources). However, it is clear from MIR spectra of galaxies (e.g. Roche *et al.* [1991]), or from narrow filter mapping (e.g. Mazzarella et al. [1994]) that emission from the interstellar bands represent a small fraction ($\approx 10\%$) of the ground-based detected emission. What probably occurs is that very close to the heating sources, grey-body emission from grains in thermal equilibrium overcomes that of impulsively heated grains.

Note that the emission detected by ground-based telescope is generally smaller than that detected by IRAS, indicating that IRAS, with its better sensitivity to extended emission, probably detected large scale diffuse emission. One can speculate that, as seems to be the case in our galaxy, much of this supplementary diffuse emission originates in small particles in quiescent clouds where the situation is reversed: large grains are too cold there to significantly emit in the MIR (see the cool dust phase section).

Besides cold dust, which is accessible by submm telescopes, the hot dust phase is the only one that can be studied from the ground in the semi-transparent N and M bands (respectively around 10 and 20 μm). As the technology just came of age in the recent years, there now exists quite a number of cameras working in this wavelength range (about 20 around the world). Although the sensitivity is limited by detector material and the atmospheric background to $\approx 1\,$Jy ($1\,\sigma$ in 1 s), the invaluable advantage of ground-based observations over IRAS or ISO is the spatial resolution: it allows to detect the detailed structure of the dusty ISM, even in distant galaxies, while resolved IRAS images of galaxies are extremely scarce.

The morphological and spectral information brought by ground-based

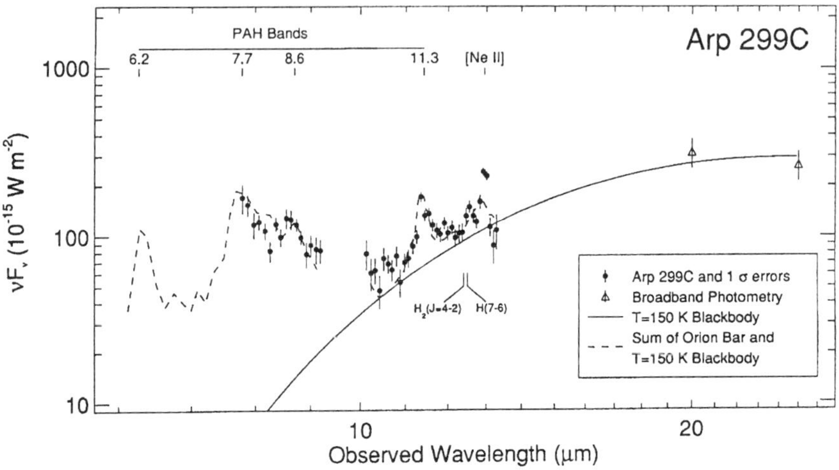

Figure 6. The MIR spectrum of Arp 299c, a component of the merging system Arp 299, is characteristic of star-forming galaxies: a hot continuum underlies the well known UIB. The black-body continuum ceases to dominate the emission shortward of 9 μm, however the latter is still dominated by the continuum associated with the UIB (figure adapted from Dudley & Wynn-Williams [1993]).

studies of galaxies (see e.g. Telesco *et al.* [1993]) both make it clear that the dust we see in the MIR is strongly associated with star formation. In fact one could say that the hot phase is the only one that could unambiguously be associated to star-forming regions.

In NGC 1097, a well-know barred galaxy with a ring of star formation, the MIR emission is concentrated in the ring, with a morphology similar to that of the CO gas, and peaks at the location of the H II regions (Telesco *et al.* [1993]).

In galaxies where they are available, the MIR and radio morphologies are also similar, though the brightnesses do not correlate well, indicating here also a common origin of the two emission (e.g. Telesco & Gezari [1992], or Keto *et al.* [1992]).

In the blue compact galaxy He 2-10, the ground-based MIR emission accounts for almost all of the emission detected by IRAS, and has a morphology very similar to that of the hot stellar clusters. Its morphology is also very reminiscent of the UV and Hα images (Sauvage *et al.* [1994]).

The MIR spectra of normal and star forming galaxies are all very similar (see Roche *et al.* [1991]) and resemble that of Galactic H II regions, i.e. they display the UIB, superimposed on a continuum that can be fitted by a hot black body ($\approx$150 K, see Figure 6).

All these observations clearly indicate the origin of the emission: in the immediate vicinity of H II regions, in the so-called photodissociation regions, where UV stellar radiation is no longer strong enough to destroy dust (either directly or by electron sputtering) or ionize hydrogen, but still hard enough to dissociate molecular clouds and heat dust to the required temperature. A recent study of the Orion H II region (Tielens et al. [1993]) has shown that the UIB emission is restricted to the photodissociation front.

As for the mass of dust that this component represents, it is difficult to estimate precisely given the uncertainties on the nature of the dust. However it is probably at best a few percent of the total dust mass (for an estimate in the framework of the coal model, see section 2.3) since it is extremely localized.

As a final note on hot dust, let me mention the fact, which is slightly out of the scope of this review, that in active galaxies (Seyfert and AGN), the MIR spectrum is markedly different, and in particular the UIB are generally absent (Cohen [1992]). This is likely related to the heating spectrum: a hard non thermal spectrum, that can profoundly affect the nature of the dust. In these galaxies, the association of MIR emission with star formation is dubious.

3.2.2. *A warm dust phase - 50 K > T_d > 25 K*

This warm dust phase is that which has the highest surface brightness in the IRAS 60 μm band when the galaxies are spatially resolved. For unresolved galaxies, one needs to know how much of the band flux is contributed by fainter extended emission from small particles such as VSGs before assessing the abundance of warm dust. Therefore most of the unambiguous results one this phase come from studies of nearby galaxies.

Contrary to the hot dust phase, this one cannot be strictly associated with star formation. Although it is true that when star formation is present, warm dust associated with it will rapidly dominate the emission, Sauvage & Thuan (1994) have shown that IRAS colors characteristic of warm dust are found in galaxies known to be devoid of star formation, i.e. normal elliptical and early-type galaxies.

To explain this result result, they pointed out that an equilibrium temperature in the range considered here requires in fact only a high enough ISRF and not necessarily the presence of hard UV photons. Such an environment is easy to find in the bulge of the latter galaxies, where the stellar density is high.

This interpretation is confirmed by the recent KAO observations of Smith *et al.* (1994) who detected FIR emission with a color temperature of $\approx$ 45 K in the bulge of quiescent early-type spirals. These authors show that the evolved stellar population present in the bulges is enough to provide the required energy.

The location of this dust phase in galaxies thus depends on the dominant origin of the heating. In early-type spirals, the emission is located in the bulge, where the stellar density is high enough. In later-type spirals, there remain traces of this warm dust component in the bulge. However, as the star formation efficiency is now high, warm dust is more abundant close to star-forming regions, in giant molecular clouds in the spiral arms of galaxies. The dominance of one of these two sites on the global FIR spectrum is more or less a function of the morphological type, i.e. of the bulge-to-disk ratio. However, as noted by Sauvage & Thuan (1994), there is in fact an intermediate state between bulge dominated galaxies and star formation dominated galaxies: intermediate-type large spirals (i.e. Sbc) with large quantities of cool dust, have a FIR spectrum that is markedly different from both elliptical and star-forming galaxies.

Given that it requires only moderately strong radiation fields, in galaxies where star formation provides most of the heating, warm dust emission is generally more extended than the Hα emission (see e.g. Sauvage *et al.* [1990], or Devereux *et al.* [1994]). A fraction of this component of the dust can be located in clouds behind the photodissociation regions, which act has a shield against hard radiation. Another fraction is probably mixed with giant molecular clouds in the vicinity of star-forming regions. However in that case, it is quite likely to be located on the surface of these clouds, as dust deeper inside should to be colder, shielded as it is from the ambient radiation field by the outer layers.

For the component heated by dense regions of old stars, it is probably physically identical to the clouds of cool dust described in the next section. What makes the difference here is only the heating field.

The relative mass fraction of this component will depend on the considered galaxy. In bulge dominated galaxies, where it remains to be demonstrated that cold dust exists, it could represent most of the dust mass. In disk galaxies however, it probably represents much less. Comparison between dust to gass mass ratio measured with IRAS on outer galaxies and that of our own Galaxy imply that IRAS only sees $\approx$10% of the dust mass. As the dust mass computed from IRAS fluxes is biased due to the presence of cool and cold dust, the fraction is probably lower.

3.2.3. *A cool dust phase - 25 K > T_d > 10 K*

Cool dust emits predominantly longward of 100 μm, though its emission will obviously be mixed with that of warm dust toward the shortest wavelengths. The very existence of this phase has long been a subject of controversy: given the uncertainties in emission properties of the grains, most of the FIR measurement (IRAS 60 and 100 μm combined with ground-based submillimetric measurement) on unresolved galaxies can be explained using a single dust temperature ($\approx$30 K) and adjusting the exponent of the emissivity law (see Eales *et al.* [1989] or Kwan & Xie [1992]).

Up to very recently, galaxies meeting our sensitivity requirements in the FIR and submm range were unfortunately selected by their high FIR fluxes, a method that is well-known to result in samples biased toward star-forming objects (e.g. Thuan & Sauvage [1992]). Given that the emission scales roughly as $T_{\rm eq}^{4+n}$, where n is the exponent of the emissivity law, cold dust is difficult to detect in these objects, if it exists at all, as the high ISRF could globally raise the temperature of all dust phases.

However data are now accumulating in the submm or from KAO on nearby spirals, where emission from star forming regions can be disentangled from that of the quiescent ISM. These data show quite unambiguously that cold dust is abundant in these galaxies. This fact had already been guessed since dust-to-gas mass ratio derived from IRAS were systematically a factor of 10 lower than that measured in our own Galaxy (Devereux & Young [1990]).

As the operating frequency of submillimeter telescopes is gradually increasing, filling up the gap toward the FIR, it is now clear that in non-starburst galaxies the peak of dust emission is located longward of 100 μm, an indisputable proof that cool dust is present. Figure 7a shows the spectrum from the central region of M 51 where cold dust emission is obvious (Chini & Krügel [1993]). Similar observations are available on a growing number of galaxies (see other galaxies in Chini & Krügel [1993], Sievers *et al.* [1994], or Franceschini & Andreani [1995]) and all show the same feature: a dust spectrum that continues to rise longward of 100 μm.

Using the very sensitive submillimetric maps that are now available on a few nearby galaxies (see e.g. Sievers *et al.* [1994], Guélin *et al.* [1993]), it is possible to study in detail the properties of this cool dust phase. It is found that this phase is closely associated with the molecular component in galaxies. In NGC 891, the brightness profiles at 1300 μm and in CO are remarkably similar (Figure 7b). In the face-on spiral NGC 3627, Sievers *et al.* (1994) find that the cool dust is concentrated in the bulge and in the spiral arms of the galaxy. The comparison of the CO interferometric map of Zhang *et al.* (1993) with the 1300 μm emission is striking: they are almost identical. Zhang *et al.* (1993) also present an H I VLA map where one can see that, contrary to CO and dust, no H I emission is detected at the center of the galaxy, while the disk is much more extended. As time goes by, new such observations confirm this trend (see Braine *et al.* [1995]).

Thus it is clear that cool dust is associated with the molecular rather than the atomic gas phase. It is also quite extended, as, in the few examples studied so far, its size is comparable to the optical size (Chini & Krügel [1993]). An obvious location for cool dust is then inside the molecular

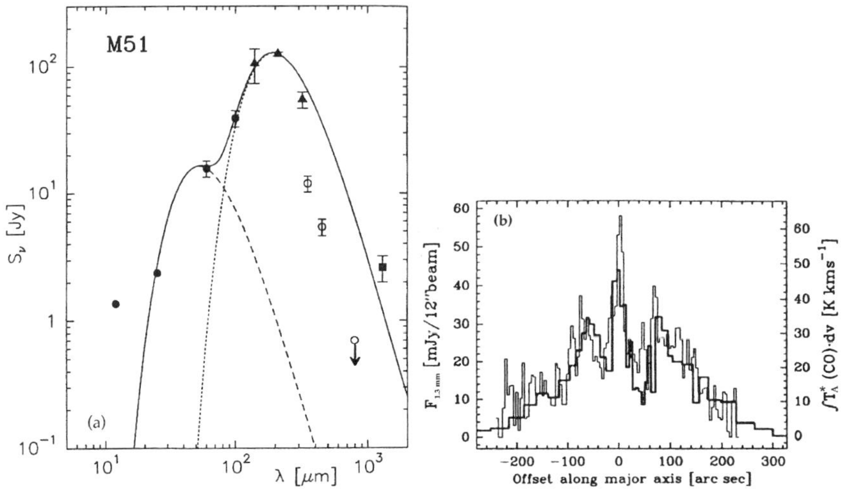

Figure 7. (a) The spectrum of the central regions of M 51 (from Chini & Krügel [1993]), open symbols represent the measurements of Eales et al. (1989). Chini & Krügel attribute the discrepancy to the too small separation between the On and Off fields of Eales *et al.* (1989), resulting in contamination of the Off field by source emission. Their data show clearly that the peak of dust emission occurs around 200 μm. Though the decomposition of emission from 25 to 1300 μm into two gray-body emissions may be too simple, the spectrum itself is evidence enough for cool dust ($T_{\rm eq} < 25$ K). (b) A comparison of the brightness profiles along the major axis of NGC 891 in the ^{12}CO (2-1) line (thick line and right-hand scale) and in the 1300 μm continuum (thick line and left-hand scale). The similarity is striking, note the coincidence in the ring position (the peaks at $\approx \pm 60$" of the center) as well as the similar scale-length of the underlying emission (figure adapted from Guélin *et al.* [1993]).

clouds. There, the outer layers of the clouds would shield them from the relatively high ISRF that bathe the spiral arms. In this picture, there is a continuity between the warm dust phase and the cool dust phase: the warm phase resides in the outer layers of the molecular clouds, or very near the newly formed hot stars, while the cool phase fills most of the volume inside these clouds. Cool dust would turn into hot dust upon cloud collapse as it would then be extremely close to newly formed stars.

This cool dust likely comprises most of the so-called infrared cirrus. Though these where thought to be predominantly associated with H I clouds exposed to the ISRF, recent studies on individual clouds in our Galaxy show a striking spatial and brightness correlation between the cirrus emission at 60 and 100 μm, and the optically thin ^{13}CO emission (e.g. Abergel *et al.* [1994]). This is completely consistent with the previously described results in nearby galaxies. It is likely however that some cirrus emission in the inner parts of a galaxy can be assigned to predominantly H I clouds (see e.g. Herter *et al.* [1990]), however it should not represent a large fraction of the cool dust mass. In the outer parts of a galaxy, where H I is predominant, the situation is probably reversed though the dust equilibrium temperature should be quite lower (see section 3.2.4).

The mass of this cool component is difficult to estimate for several reasons: (1) the uncertainty in the exponent of the emissivity law is still severe in this wavelength regime, and (2) some of the millimetric and submillimetric flux may be contributed by a still colder dust phase. However, it can already be said that in normal spirals, this component makes up most of the dust mass. Estimates based on decomposition of emission spectra result in a fraction of 80-95% of the dust being in the cool phase (Sievers *et al.* [1994], Guélin et al. [1993]).

Clearly the investigation of this dust phase will be one of the key achievements of ISO, since it has the capacity to observed in the 100-200 μm window.

3.2.4. *A cold dust phase - $T_{\rm d} \approx 10\,K$?*

This section is still speculative. There exists only very few measurements that point toward dust at such low temperature. However it would be logical to postulate that, if dust is widespread in galaxies, its temperature should reflect the various intensities of radiation field that can be found in a galaxy. In particular, at the outskirts of galaxies where the radiation field is mainly the intergalactic field, the equilibrium temperature of carbon or silicate grains is around 10 K. This temperature would be even lower in dense molecular clouds away from heating sources, though cosmic rays could then contribute an important fraction of the heating.

The problem in detecting such a dust phase is its expected low surface brightness. Cool dust is already at the limit of what current telescope can detect, cold dust, if as abundant, would be a factor of $\approx$2 dimmer.

It is only in our Galaxy that this phase can be studied as only there can large beams be used without diluting the emission. Wright *et al.* (1991), using the FIRAS instrument on COBE, show that the FIR average spectrum of the Galaxy presents an excess over what can be expected from a grey body at $T_{\rm eq} = 20$ K. This excess can be represented by a grey body emission at $\approx$5 K. It is still early to conclude that this is indisputable evidence for very cold dust, as the same spectrum can be fitted by a single gray-body temperature if one allows for flexibility in the exponent of the emissivity law. If indeed such cold dust exist, it does not represent much of the interstellar dust: Wright *et al.* (1991) conclude that the cold dust represents $\approx$ 10% of the cool dust mass.

Another obvious place to look for this very cold dust phase would be in the few H I intergalactic clouds that have been found in the Universe (see Giovanelli & Haynes [1989], Schneider *et al.* [1983], and Taylor et al. [1993]). Detection of dust in these objects would obviously put strong constraints on their origin (primordial or ejecta from galaxy collisions). Given the sensitivity required, ISO is currently the only instrument able to detect this dust.

4. The fate of dust in the ISM, and future instruments

I have abundantly discussed the spectal properties of various dust models, the emission spectrum and distribution of dust in galaxies. However there are two important points that I have not touched upon yet: the formation and processing of dust in galaxies. Though these subjects are still very much "in progress", I summarize in the following our current ideas on these two aspects.

I end this review by going through the IR range and examining what available and future instruments may help us understand.

4.1. FORMATION AND PROCESSING OF DUST GRAINS

This is by far one area where our knowledge is most fragmentary. Indeed it cumulates the disadvantages of a restricted range of observational constraints, and of uncertainties in the composition, and spectral properties of dust. I will thus only briefly summarize some of the results obtained so far.

4.1.1. *Mechanisms and sites of dust formation*

The most obvious site to look for dust formation is in the circumstellar envelopes of evolved relatively low-mass stars. Indeed in these environments we find: (1) material enriched in heavy elements such as C, O, Si..., necessary for grain formation, (2) density and temperature conditions suitable to the formation of large molecules which, through further processing, will lead to dust grains.

That dust exists in the environments of such stars is amply confirmed by IRAS observations: excess emission over the expected photospheric blackbody flux is detected toward C- and O-rich stars. Detailed inspection of the infrared spectra from these stars reveals the presence of emission and absorption bands similar to those observed in the diffuse interstellar medium: The silicate feature at 9.7 μm is present (e.g. Volk & Kwok [1987]), either in emission or absorption, as well as the unidentified infrared bands (UIB), with wavelengths slightly shifted compared to their position in the ISM. This shift can be interpreted, in the framework of the coal model for instance, as being due to the freshness of the emitting material, compared to the more processed dust in the ISM.

Another suspected site of dust formation is in supernova ejecta. Here also do we find the two conditions mentioned above: enrichment in heavy elements and suitable temperature and density conditions. So far only in SN 1987a is there some indisputable proof of the formation of dust: the monitoring of the infrared flux shows that starting at day 350, dust has been condensing in the ejecta (e.g. Meikle *et al.* [1995], Colgan *et al.* [1994]). Likely precursors of dust grains, such as SiO have also been detected in this supernova.

As can be expected, dust also probably forms during nova outbursts. Smith, Aitken & Roche (1994) have shown that features characteristic of small carbonaceous grains and silicates appeared in the mid-infrared spectrum of Nova Cen 1986.

It remains to be shown that these are the only sites of ixdust formation, and in particular whether or not dust forms in the diffuse ISM, or during the collapse of a cloud to form stars, or even in the high density clouds surrounding accretion disks. However for these latter possible sites, it would probably be more correct to speak of dust processing than of dust formation.

4.1.2. *Dust grain processing in the ISM*

Grain processing is more widely studied that grain formation, though the interpretation of observations requires the choice of a working model which weakens the strength of the conclusions. However, whatever the model chosen, processing is evident in a wide number of environments.

For instance, Bregman *et al.* (1994) have studied the relative variation of the 3.3, 8.4 and 11.4μm line in the Orion bar. Using the framework of the PAH model, they conclude that the size distribution changes within the bar, with the smallest PAHs producing most of the detected 3.3μm peaks inside the Bar, while PAHs are larger than average at the front of the Bar. Charge also changes whithin the Bar, as PAHs appear predominantly neutral along the front while they are charged inside.

Boulanger *et al.* (1990) have also shown that the abundance of small particles responsible for IRAS short wavelength emission varies inside molecular clouds. To explain these variations, they suggest that small particles are formed on larger grain surfaces inside the molecular clouds. Their abundance is then determined by the density structure of the clouds which influences the timescale of the formation, desorption and sticking processes.

Constraints on the formation and processing of dust can also be obtained on Galactic scales: Sauvage *et al.* (1990) have found that the ratio of MIR to FIR emission, that can be somehow related to the relative abundance of very small versus large dust particles, is correlated with the O/H abundance: the lower the abundance, the lower the ratio. This trend is also found in galaxies presenting a radial abundance gradient (Vigroux, *private communication*). Understanding the origin of this effect would probably shed much light on the fate of dust grains in the ISM. One could also study the dust segregation in circumstellar disks as well as the dust sputtering in interstellar shocks.

4.2. WHAT CAN BE LEARNED FROM AVAILABLE AND FUTURE INSTRUMENTS?

Since early 1996 we are receiving data from the Infrared Space Observaty ISO, a satellite able to perform imagery in the NIR and MIR ranges, photometry and spectroscopy from the NIR to the FIR. We already have in our possession 4 m telescopes equipped with MIR cameras and spectrometers, and should soon have the same instruments on 10 m telescopes. In the FIR range the KAO will be replaced by a 2.5 m airborne telescope (SOFIA), while bolometers arrays become available on larger and larger radio antennae. Thus the future looks bright for infrared astronomy. What can it bring?

4.2.1. *The mid infrared range*

On the imaging side, we will likely see a very complementary approach with ISO and ground-based telescopes, the first one bringing sensitivity, allowing to detect faint structures in nearby galaxies or high-z active galaxies, while the second, with their spatial resolution, will allow us to probe in details the star-forming regions of galaxies. This will probably tell us what role dust plays in the star forming process: does it help regulating it, or is it simply involved in it?

Sensitive spectrometers and imaging-spectrometers, either on board of ISO or on the ground, will generate a leap forward in our understanding of the dust composition. Most the proposed models are able to make predictions about the behavior of characteristic lines with density, abundances, grain processing, heating sources, ...etc. These observations will provide the long awaited tests.

4.2.2. *The far infrared range*

On the imaging side, it is likely that we will see ground and space based observatories tackling different questions: ISO with its 65 cm mirror will mostly be used for its unsurpassed sensitivity. It will allow to probe the deep universe for primeval galaxies such as F10214+4724, or study the global abundance of cold dust in galaxies, or look for possible dust in the outer H I disks of galaxies or in protogalactic H I clouds. Given the wide spectral coverage it offers, it will also be a unique instrument to study the global energetical balance in galaxies, since dust is a major agent in energy reprocessing, determine the global temperature spectrum, and study environmental and morphological effects on these properties.

Once again the main advantage of airborne observatories is the spatial resolution they offer. With a 2.5 m telescope like SOFIA, one will be able to image individual dust clouds in a large sample of galaxies, determine the distribution of the various dust phases mentioned in section 3, study their heating sources, and examine in details their links with the other phases of the ISM.

The FIR domain will probably be the one where spectrometers on board of ISO will have the larger impact, and not only because this domain is mostly out of reach of our instruments. However this impact is slightly out of the scope of this paper as it will mainly concerns the energetics of the gaseous ISM through the study of the numerous fine structure lines present in the FIR. Still these instruments will allow for the first time high-resolution spectra to be performed on the continuum FIR emission of galaxies. This will allow us to precisely determine the distribution of dust temperature instead of the simplistic multi-component models in use so far (involving the one exposed in section 3). This distribution and its variations will be extremely valuable to characterize the connections between dust, the rest of the ISM, and the stellar component.

4.2.3. *The submillimeter and millimeter range*

This is a domain where much is to be expected from ground-based telescopes: as the bolometer technology comes of age, we will see larger and larger arrays of increased sensitivity installed on large radio antennae. From the results already obtained by such arrays either with the 30 m IRAM telescope, or the 15 m JCMT telescope on nearby galaxies, one can draw the broad lines of future research: determination of the absorption coefficient of dust at long wavelength through a comparison of observations and models, and detailed study of the physics of the cool and cold dust phase in galaxies. In particular it would be interesting to obtain precise estimates on the mass of these components and the abundance of dust in cold neutral H I clouds. These informations would probably give us clues on such phenomena as formation, processing, diffusion and survival of dust in the ISM of galaxies as well as shed some light on the formation of galaxies themselves, if it turns out that dust is abundant even in the outskirts of galaxies.

Obviously another step forward in all these areas has to be expected from FIRST, however it is preferable to leave that to an equivalent review one year before its expected launch in the years 2005...

Though it is not my intention (neither is it in my reach) to give an exhaustive list of what the future will bring, the aspects mentioned above already show that we have ahead of us a decade of hard but exciting work. Let us hope that in a few years from now we will have a brand new infrared view of our Universe, glittering with sources we do not even expect yet.

5. Conclusion

I have attempted to summarize our current knowledge on the properties of dust in the disk of galaxies, even before early results are available from ISO, the satellite which will probably bring some revolutionary information to the field. This will be a good starting point for understanding new observations from ISO and other facilities.

The main conclusions can be summarized as follows:

• **Dust Models.** It is important to remember that so far, it is not possible to tell which among the various proposed models, corresponds most closely to dust in galaxies. Though it is agreed that interstellar dust should consist mainly of silicates and carbonaceous grains, this still leaves room for large uncertainties that have their impact on properties deduced from observations: (1) the color temperature can only be used as an indication of the thermodynamical state of the dust. In a realistic situation, different phases and species of dust grains are present on any given line of sight, resulting in an observed temperature that has probably little connection with any physical temperature. (2) The dust mass derived from the observations is also at best an estimate of the total dust mass: cold dust is hard to detect with IRAS though it will affect the mass determination. At longer wavelengths, the uncertainty on the spectral index of the emissivity law also results in large uncertainties, but the total dust mass is better probed.

In general, thus, a comparison of observed properties with models without taking into account the limitations brought by the instruments or the uncertainties from the model themselves is bound to produce spurious results.

• **Heating Sources.** Star formation, and the high-mass ionizing stars that are created in the process, is obviously an important provider of energy to the dust phase. Little star formation activity is enough in fact to provide the dominant heating flux in a galaxy. However high far infrared luminosities are observed in galaxies with very low levels or no star formation at all. A careful examination of the energetical balance in galaxies show that in fact, the radiation field from older stars also provides some heating to the dust phase. It can be shown that this represents the dominant source either for early-type galaxies or for selected regions of later-type galaxies.

The fact that more that one source of heating has to be considered is important as it implies for instance that the infrared luminosity cannot be used straightforwardly as a star formation indicator.

• **Dust phases.** Connected to the previous point, it appears clearly now that dust is present throughout the ISM of a galaxy and not only in its densest star forming regions. As a result a broad spectrum of temperatures is detected. These different temperatures can be related to dust in different regions of the ISM: the hottest dust phase is directly associated with the star forming regions, close to the photodissociation fronts created by the high mass stars. Warm clouds of dust are probably also linked to star formation though they are likely to be more extended than the previous phase. They are probably located on the outer layers of molecular clouds close to star forming regions. Note that warm dust in present in bulge dominated galaxies that is not at all related to star formation. In these galaxies it is probably diffuse. Cool dust can be found inside molecular clouds and is no longer related to star formation. Finally cold dust probably exist on the outskirts of galaxies, in diffuse H I clouds or in very dense molecular clouds away from heating source. It can be postulated that the relative abundance of a dust phase is inversely proportional to its temperature though only future work can precisely assess this hypothesis.

Acknowledgements

Section 2.1 greatly benefited from an enlightening seminar by L. Colangeli. I acknowledge the help of R. Papoular for pointing out the strengths and weaknesses of various dust models. A large part of the material for this paper was drawn from a long-standing collaboration with T.X. Thuan and L. Vigroux. I benefited from the insights of G. Helou, and J. Lequeux on the dusty ISM of galaxies. J. Mathis and J. Bregman provided me with very valuable input for this review.

References

Abergel, A., Boulanger, F., Mizuno, A. and Fukui. Y., 1994, *Ap.J.* , 423, L59.
Boulanger, F., Falgarone, E., Puget, J.-L. and Helou, G., 1990, *Ap.J.* , 364, 136.
Braine, J., Kruegel, E., Sievers, A. and Wielebinski, R., 1995, *Ast.Ap.* , 295, 55.
Bregman, J., Larson, K., Rank, D. and Temi, P., 1994, *Ap.J.* , 423, 326.
Bregman, J.D., Allamandola, L.J., A.G.G.M. Tielens, Geballe, T.R. and Witteborn, F.C. 1989, *Ap.J.* , 344, 791.
Chini, R. and Krügel, E., 1993, *Ast.Ap.* , 279, 385.
Cohen, M., 1992, *A.J.* , 103, 6.
Colgan, S.W., Haas, M.R., Erickson, E.F., Lord, S.D. and Hollenbach, D.J. , 1994, *Ap.J.* , 427, 874.
Condon, J.J., Anderson, M.L. and Helou, G., 1991, *Ap.J.* , 376, 95.
Cox, P. and Mezger, P.G. 1989, *Ast.Ap.Rev.* , 1, 49.
Devereux, N.A. and Young, J.S., 1990, *Ap.J.* , 359, 42.
Devereux, N.A. and Young, J.S., 1991, *Ap.J.* , 371, 515.
Devereux, N.A., Price, R., Wells, L.A. and Duric, N., 1994, *A.J.* , 108, 1667.
Désert, F.-X., Boulanger, F. and Puget, J.-L. 1990, *Ast.Ap.* , 237, 215.
Draine, B.T. 1981, *Ap.J.* , 245, 880.
Draine, B.T., 1990, in *The interstellar medium in galaxies*, eds. H.A. Thronson, Jr. and J.M. Shull, Kluwer Academic Publisher, The Netherlands, p. 483.
Draine, B.T. and Lee, H.M. 1984, *Ap.J.* , 285, 89.
Dudley, C.C. and Wynn-Williams, C.G., 1993, *Ap.J.(Lett.)* , 407, 65.
Duley, W.W. and Williams, D.A. 1988, *M.N.R.A.S.* , 231, 969.

Duley, W.W. and Williams, D.A., 1994 *M.N.R.A.S.* , 272, 441.
Eales, S.A, Wynn-Williams, C.G. and Duncan, W.D. 1989, *Ap.J.* , 339, 859.
Engargiola, G., 1991, *Ap.J.Suppl.* , 76, 875.
Engargiola, G. and Harper, D.A., 1992, *Ap.J.* , 394, 104.
Franceschini, A. and Andreani, P., 1995, *Ap.J.(Lett.)* , 440, L5.
Giard, M., P, F., Lamarre, J.M., Serra, G. and Caux, E. 1989, *Ast.Ap.* , 215, 92.
Giovanelli, R. and Haynes, M.P. 1989, *Ap.J.(Lett.)* , 346, L5.
Guélin, M., Zylka, R., Mezger, P.G., Haslam, C.G.T., Kreysa, H., Lemke, R. and Sievers, A.W., 1993, *Ast.Ap.* , 279, L37.
Guillois, O., Nenner, I., Papoular, R. and Reynaud, C. 1994a, in *Molecules and grains in space*, ed. I. Nenner, AIP Press, New-York, p. 817.
Guillois, O., Nenner, I., Papoular, R. and Reynaud, C., 1994b, *Ast.Ap.* , 285, 1003.
Helou, G. 1986, *Ap.J.(Lett.)* , 311, L33.
Herter, T., Shupe, D.L. and Chernoff, D.F., 1990, *Ap.J.* , 352, 149.
Jones, A.P., Duley, W.W. and Williams, D.A. 1987, *M.N.R.A.S.* , 219, 213.
Keto, E., Ball, R., Arens, J., Jernigan, G. and Meixner, M., 1992, *Ap.J.* , 389, 223.
Koornneef, J. and Code, A.D. 1981, *Ap.J.* , 247, 860.
Kwan, J. and Xie, S., 1992, *Ap.J.* , 398, 105.
Lee, W. and Wdowiak, T.J., 1993, *Ap.J.* . Lett., 410, L127.
Léger, A. and Puget, J.-L. 1984, *Ast.Ap.* , 137, L5.
Leisawitz, D. and Hauser, M.G. 1988, *Ap.J.* , 332, 954.
Mathis, J.S., 1990, *Ann.Rev.Ast.Ap.* , 28, 37.
Mathis, J.S., 1994, *Ap.J.* , 422, 176.
Mathis, J.S. and Cardelli, J.A., 1992, *Ap.J.* , 368, 610.
Mathis, J.S. and Whiffen, G. 1989, *Ap.J.* , 341, 808.
Mathis, J.S., Metzger, P.G. and Panagia, N. 1983, *Ast.Ap.* , 128, 212.
Mazzarella, J.M., Voit, G.M., Soifer, B.T., et al., 1994, *A.J.* , 107, 1274.
Meikle, W.P.S., Spyromilio, J., Allen, D.A., Varani, G.F. and Cumming, R.J., 1993, *M.N.R.A.S.* , 261, 535.
Papoular, R., Reynaud, C. and Nenner, I., 1991, *Ast.Ap.* , 247, 215.
Papoular, R., Breton, J., Gensterblum, G., Nenner, I., Papoular, R.J. and Pireaux, J.-J., 1993, *Ast.Ap.* , 270, L5.
Papoular, R., Guillois, O., Nenner, I. and Reynaud, C., 1995, *Ast.Ap.* , 293, 562.
Phillips, M.M., Aitken, D.K. and Roche, P.F. 1984, *M.N.R.A.S.* , 207, 25.
Prévot, M.L., Lequeux, J., Maurice, E. Prévot, L. and Rocca-Volmerange, B. 1984, *Ast.Ap.* , 132, 389.
Puget, J.-L. and Léger, A. 1989, *Ann.Rev.Ast.Ap.* , 27, 37.
Puget, J.-L., Léger, A. and d'Hendecourt, L., 1995, *Ast.Ap.* , 293, 559.
Rand, R.J., Kulkarni, S.R. and Rice, W., 1992, *Ap.J.* , 390 66.
Roche, P.F., Aitken, D.K., Smith, C.H. and Ward, M.J., 1991, *M.N.R.A.S.* , 248, 606.
Rowan-Robinson, M., Broadhurst, T., Lawrence, A., *et al.* , 1991, *Nature* , 351, 719.
Russel, R.W., Soifer, B.T. and Willner, S.P. 1977, *Ap.J.(Lett.)* , 217, L149.
Sakata, A., Wada, S., Onaka, T. and Tokunaga, A.T. 1987, *Ap.J.(Lett.)* , 320, L63.
Sauvage, M. and Thuan, T.X., 1992, *Ap.J.(Lett.)* , 396, L69.
Sauvage, M. and Thuan, T.X., 1994, *Ap.J.* , 429, 153.
Sauvage, M., Lagage, P.O. and Thuan, T.X., 1994, in *ESO-OHP workshop on dwarf galaxies*, eds. G. Meylan and P. Prugniel, ESO, Garching, p. 443.
Sauvage, M., Thuan, T.X. and Vigroux, L., 1990, *Ast.Ap.* , 237, 296.
Savage, B.D. and Mathis, J.S. 1979, *Ann.Rev.Ast.Ap.* , 17, 73.
Schlemmer, S., Cook, D.J., Harrison, J.A., Wurfel, B., Chapman, W. and Saykally, R.J., 1994, *Science* , 265, 1686.
Schneider, S.E., Helou, G., Salpeter, E.E. and Terzian, Y. 1983, *Ap.J.(Lett.)* , 273, L15.
Sellgren, K. 1985, *Ap.J.* , 277, 623.
Siebenmorgen, R., 1994, *Ast.Ap.* , 282, 407.
Siebenmorgen, R. and Krügel, E., 1992, *Ast.Ap.* , 259, 614.

Sievers, A.W., Reuter, H.-P., Haslam, C.G.T., Kreysa, H. and Lemke, R., 1994, *Ast.Ap.* , 281, 681.
Smith, B.J., Harvey, P.M., Colomé, C., Zhang, C.Y. and DiFrancesco, J., Pogge, R.W., 1994, *Ap.J.* , 425, 91.
Smith, B.J., Harvey, P.M. and Lester, D.F., 1995, *Ap.J.* , 442, 610.
Smith, C.H., Aitken, D.K. and Roche, P.F., 1994, *M.N.R.A.S.* , 267, 225.
Taylor, C.L., Brinks, E. and Skillman, E.D., 1993, *A.J.* , 105, 128.
Telesco, C.L. and Gezari, D.Y., 1992, *Ap.J.* , 395, 461.
Telesco, C.M., Dressel, L.L. and Wolstencroft, R.D., 1993, *Ap.J.* , 414, 120.
Thuan, T.X. and Sauvage, M., 1992, *Ast.Ap.* . Suppl., 92, 749.
Tielens, A. G. G. M., Meixner, M. M., van der Werf, P. P., Bregman, J., Tauber, J. A. *et al.* , 1993, *Science* , 262, 86.
Volk, K. and Kwok, S. 1987, *Ap.J.* , 315, 654.
Völk, H.J. and Xu, C., 1994, Infrared Phys. Technol., 35, 527.
Whittet, D.B. 1988, in *Dust in the Universe* eds. M.E. Bailey & D.A. Williams, Cambridge University Press, p. 25.
Wilson, C.D., Scoville, N. and Rice, W., 1991, *A.J.* , 101, 1293.
Wright, E.L., Mather, J.C., Bennet, C.L., Cheng, E.S., Shafer, R.A., *et al.* , 1991, *Ap.J.* , 381, 200.
Zhang, X., Wright, M. and Alexander, P., 1993, *Ap.J.* , 418, 100.

MOLECULAR GAS IN GALAXY DISKS

JEFFREY KENNEY
Yale University
Astronomy Department
P.O. Box 208101
New Haven, CT 06520-8101
U.S.A.

1. Introduction

From the frigid outer extremes of galaxies to their seething, energetic cores, molecular gas plays a prominent role in the structure and evolution of galaxies. molecular gas comprises the coldest and thinnest disk component in galaxies. It is the star forming part of the interstellar medium, and in some cases is a significant mass component within the galaxy. The evolution of galaxies is partly driven by the radial redistribution of matter, and the matter which flows into the central kpc of galaxies is predominantly molecular gas. molecular gas disks on 0.1–few 100 pc scales are believed to fuel starbursts as well as obscure active galactic nuclei (AGN). There is recent discussion over whether the amount of molecular gas in the outer disks of galaxies has been greatly underestimated, perhaps even to the extent that it might be the dark matter in galaxies.

In this chapter I review recent results on molecular gas in galaxies, focussing primarily on the gas distribution and kinematics in relatively undisturbed galaxies. What molecular gas teaches us about the structure and evolution of disk galaxies is a major emphasis of this paper. There are a number of good reviews on closely related topics. Young & Scoville (1991) provide a general review on molecular gas in galaxies. Henkel, Baan & Mauersberger (1991) and Mauersberger & Henkel (1993) review dense gas in the central regions of galaxies. Combes (1991) describe molecular gas throughout the Milky Way, while Blitz *et al.* (1993) paint a coherent picture of the ISM in the center of the Milky Way. Wilson (1994) discusses molecular gas properties in Local Group galaxies. Scoville *et al.* (1994)

J. M. van der Hulst (ed.), The Interstellar Medium in Galaxies, 33–74.

review atomic and molecular gas in luminous interacting galaxies. I shall emphasize topics which these people have not.

The interesting and observationally addressable questions on molecular gas vary strongly with galactocentric radius, and this chapter is organized accordingly. In the distant outer disks the quantity of molecular gas is debated (§2). For most undisturbed spiral galaxies, the mid-disks (~1–10 kpc) contain gaseous spiral structure, whose character differs between grand design and flocculent spiral galaxies (§4). The inner disks (~1 kpc) of even weakly barred galaxies contain elongated gaseous features with large non-circular streaming motions, caused by stellar bars or loosely wound stellar or gaseous spiral arms (§5). Circumnuclear regions (~100 pc) contain a variety of gaseous morphologies including rings or partial rings, miniature spiral arms, "twin peaks", or filled exponential disks (§6). In many galaxies these large gas concentrations lie near inner Lindblad resonances. Circumnuclear regions are the home of starbursts, compact circumnuclear disks, and the dense central parts of bulges. The behaviour of gas here is complex, and the molecular gas morphology and physical state of the ISM are probably strongly time-dependent. Among the most exciting new work on molecular gas concerns the inner extremes of galaxies (§8). A rotating molecular disk with a radius of ~0.1 pc detected around the active nucleus in NGC 4258 not only provides the best dynamical evidence to date of a nuclear black hole, but is probably the inner edge of a molecular torus. Molecular tori are believed to obscure the nuclei and broad-line regions of Seyfert 2 galaxies. Surprisingly dense molecular gas on 100 pc scales in the Seyfert 2 galaxy NGC 1068 may owe its extreme physical properties to energy input from the active nucleus.

There are topics other than these included in this chapter, and it seems only fair to provide an brief outline. §2 discusses molecular mass determinations. §3 describes the H I /H_2 ratio in galaxies. §4–6, which are the heart of this paper, discuss the distribution of molecular gas in the disk plane: §4 covers spiral structure, §5 covers bars, and §6 covers circumnuclear regions, including starbursts. §7 discusses the vertical distribution of molecular gas, and §8 describes new results on molecular gas around AGN.

2. How well can we measure the mass of molecular gas?

molecular gas is dominant form of gas mass in the inner parts of many galaxies, and perhaps is the dominant form of mass in centers of some galaxies (Turner 1994; Scoville *et al.* 1994). It is clearly important to accurately estimate the mass of molecular gas in galaxies. Here, we give a brief overview of the basis for molecular gas mass determinations from the CO line, a topic which is covered more thoroughly in many of the review papers cited in the Introduction. We also describe in more detail recent work on a couple of problem areas.

2.1. THE CO–H_2 RELATIONSHIP IN THE MILKY WAY AND A FEW OTHER PLACES

The bright CO(1-0) emission line is a fairly reliable H_2 mass tracer on large scales in some environments, and is the most widely used tracer of cold H_2 mass. Three main techniques are used to estimate the masses of molecular clouds and thereby calibrate the CO–H_2 relationship within the Milky Way.

1. The virial theorem.
molecular clouds can be characterized by their peak CO brightness temperature T_b, linewidth ΔV, radius R, and luminosity $L_{CO} = \pi R^2 T_b \Delta V$. The total (virial) mass of a self-gravitating object is $M_{vir} \simeq R\Delta V^2/G$. If clouds are near virial equilibrium, then the CO luminosities and virial masses of uniform clouds are related by $M_{vir} \propto (\rho^{1/2}/T_b) L_{CO}$, where ρ is the cloud density (e.g., Young & Scoville 1991). The good correlation between CO luminosities and computed "virial masses" of Milky Way clouds shows that clouds are consistent with being self-gravitating (Solomon *et al.* 1987). By itself, this correlation does not prove that clouds are self-gravitating, since the size-linewidth relation ($\Delta V \propto R^{0.5}$, Larson 1981) which exists for clouds will tend to produce a correlation between L_{CO} and $R\Delta V^2/G$. It is the agreement of the virial mass estimates with other mass estimates which shows that clouds are self-gravitating and that CO is a reasonably good mass tracer .

2. Gamma ray emissivity.
Most of the diffuse galactic gamma-ray emission originates from the interaction of cosmic rays with gas particles. A comparison of gamma ray, CO and H I maps, and the assumption that the cosmic ray density is not enhanced inside molecular clouds, leads to a fairly constant ratio of H_2 column density N_{H_2} to integrated CO temperature $I_{CO} = \int T_b dv$ for Milky Way clouds (Strong *et al.* 1988; Bloemen 1989).

3.) Optically thin molecular lines and dust extinction. Measurements of optically thin lines (e.g., $C^{18}O$), can also be used to estimate H_2 column densities, if the gas temperature is known and the relative abundance of the molecule with respect to H_2 is assumed, or if the dust extinction is measured and the gas-to-dust ratio is assumed.

The good agreement between these techniques in nearby giant molecular clouds demonstrates that molecular clouds with diameters of 3–50 pc and masses of 10^4–3×10^6 $M_\odot$ are indeed self-gravitating and in approximate virial equilibrium. Most determinations of the CO–H_2 relationship for large Milky Way clouds are consistent to within a factor of 2, and many people refer to a "standard value" which is generally in the range N_{H_2}/I_{CO} = $3\pm1\times10^{20}$ molecules cm^{-2} (K km $s^{-1})^{-1}$.

There is molecular gas known in the Milky Way which is not in virial equilibrium. Translucent or diffuse clouds, often observed at high latitude in the Milky Way, contain too little mass to be self-gravitating (Magnani, Blitz, & Mundy 1985). It is difficult but important to determine the CO–H_2 relationship in these diffuse clouds, since diffuse gas may contribute significantly to the overall CO emission from the Milky Way (Polk *et al.* 1988). Magnani & Onello (1995) find that there may be large variations in the CO–H_2 relationship among diffuse clouds, although the average value is close to the standard value.

Both the virial theorem technique and the optically thin lines technique have been directly applied in other galaxies. For galaxies where no direct measurement is available, the virial theorem offers guidance on how the CO–H_2 relationship might be expected to vary. Assuming that CO is optically thick over the entire H_2 cloud, and that the cloud is self-gravitating and in approximate virial equilibrium, the CO–H_2 relationship depends on $\rho^{1/2}/T_g$, where ρ is the gas density and T_g is the temperature (Dickman, Snell, & Schloerb 1986). It should be emphasized that this relationship does not hold along a given line-of-sight (CO is assumed to be optically thick!), but only if the entire self-gravitating cloud is within the telescope beam. Although $\rho^{1/2}/T_g$ certainly varies throughout the universe, densities are correlated with temperatures in some cases, so that $\rho^{1/2}/T_g$ varies less than ρ or T_g alone. In the central part of the starburst galaxy M82, the gas is both hotter and denser, so that $\rho^{1/2}/T$ is close to the value for local Milky Way clouds (Young & Scoville 1991), although this density-temperature compensation is not perfect and the 'standard' CO–H_2 relationship may overestimate the amount of H_2 in M82 by factor of 2–3 (Wild *et al.* 1992).

The standard CO–H_2 relationship does work well in the Milky Way over radii $2\leq R\leq8$ kpc despite a metallicity gradient. Gamma ray emission is notably weak in the central 400 pc (Strong *et al.* 1988; Bloemen 1989), indicating that the CO–H_2 relationship is different there, or that a poloidal magnetic fields shepards relativistic protons from the center (Blitz *et al.* 1993). It applies in the inner part of M33 (Wilson & Scoville 1990, 1992), and the star-forming ring of M31 (Wilson & Rudolph 1993), but may overestimate H_2 masses by a factor of 2–3 in the spiral arms of M51 (Adler *et al.* 1992; Garcia-Burillo *et al.* 1993a; Rand 1993a; see §4.1). It may signficantly underestimate H_2 masses in the Magellanic Clouds (Rubio, Lequeux & Boulanger 1993) and other low-mass galaxies (see §2.2). Its value has been questioned in starbursts and ultraluminous galaxies (Maloney & Black 1988; Wild *et al.* 1992; Shier, Rieke, & Rieke 1994), the outer parts of disks (Pfenniger, Combes, & Martinet 1994; see §2.3) and the cold parts of disks (Allen & Lequeux 1993). It will be important to quantify the CO–H_2 relationship in the environmental extremes of metallicity, pressure, and ultraviolet and cosmic ray heating.

Observations of different molecules and transitions are extremely useful for determining gas temperatures, gas densities, cloud geometries, and improving H_2 mass estimates. Ratios of ^{12}CO lines with those from the rarer isotopes ^{13}CO and $C^{18}O$, which are more optically thin than ^{12}CO, are useful for determining the optical depth of the lines, gas densities, and excitation temperatures. HCN, HCO^+, and CS require high densities for collisional excitation, and therefore trace gas with densities $n \geq 10^5$ cm^{-3}. In contrast, CO has a small dipole moment and therefore is collisionally excited at densities $n \geq 10^3$ cm^{-3}, which is the density at which most of the gas mass exists in nearby Milky Way clouds. This is part of the reason why CO is a useful probe of the H_2 in molecular clouds.

2.2. THE MAGELLANIC CLOUDS AND OTHER LOW MASS GALAXIES

CO emission is weak in irregular and other low mass galaxies. The clearest insight into the reason for this comes from recent survey results on the Small Magellanic Cloud (SMC). Rubio, Lequeux & Boulanger (1993) mapped a small region of the SMC in the (1-0) and (2-1) lines of both CO and ^{13}CO, and as shown in Figure 1, find that clouds in the SMC follow the same size-linewidth relation as the as local Milky Way (i.e., $\Delta V \propto R^{0.5}$, Larson 1981). Furthermore, for a given linewidth or virial mass, the *smallest* SMC clouds (10 pc) have about the same CO luminosity, within a factor of ~2–3, as their counterparts in the Milky Way. However, the *largest* SMC complexes (200 pc) have much lower CO luminosities, for a given linewidth or virial mass, than their counterparts in the Milky Way. What can explain this cloud size dependence? Line ratios show that the CO clouds are warm and clumpy, and have a much lower filling factor than local Milky Way clouds. CO is found only in the denser parts of the SMC clouds, which explains why the CO luminosities of large SMC clouds are comparatively low. The low CO luminosities result from the combined effects of lower C and O abundances (5 and 20 times lower than local Milky Way), a lower dust-to-gas ratio (10 times lower), and a stronger UV radiation field (10 times higher) (Israel *et al.* 1993). A cloud size dependence to the CO–H_2 relationship in small galaxies, remarkably consistent with the SMC observations, was predicted by Elmegreen (1989).

It is not certain that the large (~100 pc) SMC complexes are in virial equilibrium, but if so, the relation between CO line intensity and the proton column density $N(H_2+2H)$ is ~10 times lower on large scales (200 pc) in the SMC than the local MW. Part of this difference may be due to a higher

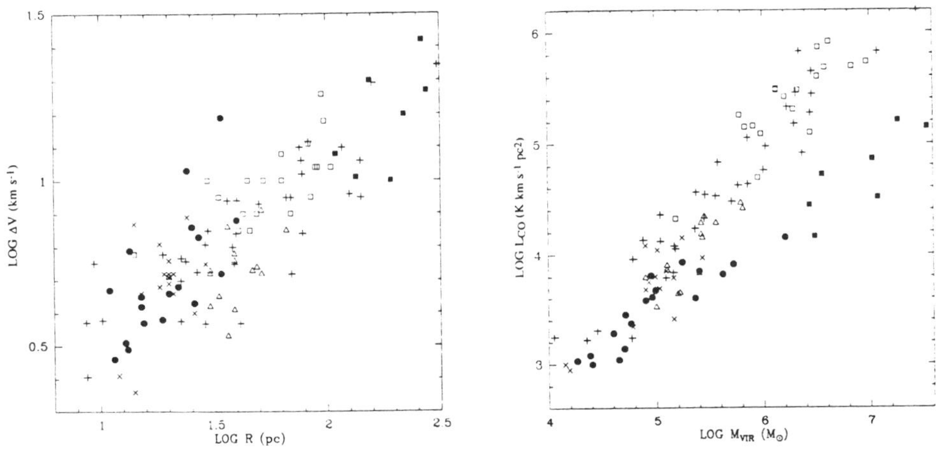

Figure 1. A comparison of CO molecular cloud properties in the Small Magellanic Cloud and the Milky Way. a.) CO(1-0) line widths ΔV versus radius R. b.) CO(1-0) luminosities $L_{\rm CO}$ versus virial masses $M_{\rm vir}$. Solid circles and solid squares represent SMC clouds, oblique crosses, open triangles, and straight crosses represent outer disk Milky Way clouds, and open squares represent inner disk Milky Way clouds. Panel (a) shows that the SMC and the Milky Way clouds obey the same size-linewidth relationship, while panel (b) shows that the CO luminosities of the largest SMC clouds are low with respect to their linewidths or virial masses. From Rubio *et al.* (1993).

atomic gas mass fraction for SMC clouds, since the stronger UV radiation field in the SMC can photodissociate a larger column density of H_2, and the low dust-to-gas ratio implies that it is harder to form H_2 in the first place (Lequeux *et al.* 1994). It will be important to determine the atomic gas fraction and therefore the CO–H_2 relationship for these large SMC complexes. Even if the CO–H_2 relationship for these large SMC complexes turns out to be similar to the Milky Way value because the clouds are H I -dominated, there could still be a problem determining cloud gas masses in other galaxies. Is H I optically thick in these clouds? If so, the standard assumption of optically thin H I will yield an underestimate of their atomic gas mass.

Low mass spiral galaxies also have weak CO emission. The ratio of CO to blue luminosities drops significantly for Virgo Sc galaxies fainter than absolute magnitude M_B = -19 (Kenney & Young 1988), suggesting that the phenomenon of weak CO emission is associated with low mass galaxies, not just irregular galaxies. Weak CO emission in small galaxies may be due to low metallicities, although there should be little variation in the CO–H_2 relationship once the metallicity is sufficiently high that optically thick CO covers most of the H_2-dominated part of a cloud. Weak CO emission may also be partly due to a lower H_2/H I ratio, since a larger fraction of the ISM in low mass galaxies is expected to be in the atomic phase due to low dust-to-gas ratio and low pressure (Elmegreen 1993).

2.3. COLD MOLECULAR GAS AND DARK MATTER

CO emission becomes weak in the outer extremes of galaxies. Since cloud heating and the metal abundance both diminish with increasing radius (on average), there is concern that the standard CO–H_2 value may underestimate the amount of molecular gas in the outer parts of galaxies. There are suggestions that clouds in the outer Milky Way have somewhat lower CO emission line luminosities for given a size or linewidth than clouds in the inner galaxy (Mead & Kutner 1988; Sodroski 1991; Digel, de Geus & Thaddeus 1994), although the results are not as clearcut as in the Magellanic Clouds, and the maximum possible difference is not large enough to make H_2 the dominant form of gas in the outer galaxy.

Emission line studies could miss colder gas because of low emissivity or a search bias, so CO absorption line searches in the outer galaxy are valuable. Lequeux, Allen, & Guilloteau (1993) detected CO absorption towards 2 quasars, and although they cannot determine the CO excitation temperature or gas column density without making assumptions, they suggest that the gas may be cold ($\sim$3.5 K) and the column density large enough such that H_2 is the dominant form of gas at R = 12 kpc. Emphasizing the large uncertainty of the assumptions, and contributing new CO and HCO^+ absorption line detections and non-detections, de Geus & Phillips (1996) conclude that it is too early to tell whether there is much cold H_2 in the outer Milky Way.

Although the outer galaxy is similar to the Magellanic Clouds in terms of low metallicity, the environment is very different in terms of heating. Due to the very different star formation rates, the UV radiation field is weak in the outer galaxy, but very strong in the Magellanic Clouds. An environment of high metallicity and low heating is the inner disk of M31, inside the well-known gas-rich annulus of active star formation at 8–11 kpc (Dame *et al.* 1993). Allen & Lequeux (1993) have detected CO from dust clouds in this inner disk of M31, and suggest that the standard CO–H_2

value may underestimate the amount of molecular gas there.

Could the dark matter in galaxies be cold molecular gas in outer disks? This question is thoughtfully explored in a recent pair of papers by Pfenniger, Combes, & Martinet (1994) and Pfenniger & Combes (1994) [see also chapter 3, p. 88]. Their provocative hypothesis is motivated by the fact that the most significant dark matter problem in galaxies is in their outer parts, and the fact that the inferred radial distribution of dark matter is similar to the observed radial distribution of H I (Bosma 1981; Freeman 1992; Broeils 1992). If the H I mass in outer galaxies were multiplied by a factor of ~20, this could solve many aspects of the dark matter puzzle in galaxies. Pfenniger, Combes, & Martinet (1994) propose that the dark matter is cold (~3K) H_2 gas with a fractal structure, in a rotationally supported disk which is coextensive with the observed H I . Located far from sites of star formation, gas in the cold outer disk might indeed have a temperature close to the 2.7K background radiation temperature, and be difficult to directly detect.

By adopting the fractal dimension of known molecular clouds suggested by self-similar isophotal cloud boundaries (Falgarone *et al.* 1992), and the Larson (1981) scaling relations between cloud sizes, linewidths, and masses, and extrapolating these to very small scales, Pfenniger & Combes (1994) suggest that most of a cloud's mass may be in very small dense lumps (as small as $R \sim 30$ AU, $M \sim 1$ Jupiter mass). If such dense lumps existed in the outer galaxy, they would indeed be hard to detect. In addition to being cold (near 3K), they would be optically thick in H I , composed mostly of H_2, and have a low metallicity so would not be detectable by absorption lines of CO or other molecular tracers.

This proposal raises many questions. If such dense gas lumps form in the outer galaxy, why don't stars form in them? Why should the radial distribution of molecular gas be the same as H I in the outer galaxy but very different in the inner galaxy (see §3). Are dark matter haloes really disks? Of course, it is easy to criticize nearly every theory of dark matter!

3. H_2 vs. H I

molecular gas is the dominant form of the ISM in the centers of gas-rich disks, exceeding the amount of atomic gas by as much as 4 orders of magnitude in extreme cases. Peak H_2 surface densities inferred from CO maps with the standard CO–H_2 conversion are typically 10–500 $M_\odot$ pc^{-2} in normal luminous spiral galaxies (Young & Scoville 1991; Kenney & Young 1989), 10^3–10^4 $M_\odot$ pc^{-2} in the centers of starburst galaxies (Kenney, Carlstrom & Young 1993), and reach values as high as 10^5 $M_\odot$ pc^{-2} in the centers of merging, ultraluminous galaxies like Arp 220 (Scoville *et al.* 1991). Azimuthally-averaged H I surface densities rarely exceed 10 $M_\odot$ pc^{-2}, although are not well known in central regions, where H I absorption is common (e.g., Koribalski, Dickey, & Mebold 1993), and could be underestimated throughout disks since H I is generally assumed to be optically thin. CO radial profiles are generally strongly centrally peaked, while H I radial profiles are relatively flat over most of the optical disk, so that the H_2/H I ratio decreases strongly with radius in most galaxy disks, and H I is probably the dominant form of the ISM in outer disks (although see §2.3).

Under what conditions does gas become molecular? Not all H I clouds are diffuse, nor are all H_2 clouds self-gravitating. There are diffuse clouds which are principally molecular, as well as self-gravitating clouds which are principally atomic. The physical conditions which govern the molecular fraction and the gravitational boundedness are discussed by Elmegreen (1993).

The column density $N_{\rm HI}$ of the H I envelope surrounding a molecular core is determined by the balance between H_2 formation on dust grains, and destruction by photodissociation, and is approximately proportional to:

$$N_{\rm HI} \propto \left(\frac{n_{\rm gas} R_{\rm H}}{G_{\rm surf}}\right)^{-1.5}$$

where $n_{\rm gas}$ is the mean gas density in the transition region, $G_{\rm surf}$ is the H_2 photodissociation rate at the cloud surface and $R_{\rm H}$ is the rate constant for H_2 formation on dust grains (Federman, Glassgold, & Kwan 1979; Elmegreen 1993). In the central parts of galaxies, where the azimuthally averaged gas surface density is high, and where the interstellar pressure is high enough to compress cloud envelopes, the gas density $n_{\rm gas}$ is sufficiently high that the H I envelope is small. This is the principle reason why the H_2/H I ratio varies so dramatically with radius in galaxies. High pressures in the intercloud medium, which compress molecular cloud envelopes, may also explain why HCN/CO and CS/CO ratios are systematically higher in galaxy centers (Rieu *et al.* 1992; Helfer & Blitz 1993).

The molecular fraction appears higher in high-mass galaxies than low-mass galaxies (Kenney & Young 1988; Elmegreen 1989), and higher in earlier type spirals than late types (Young & Knezek 1989). These effects are much smaller than the radial variation in H_2/H I observed in most galaxy disks, but may also be due to pressure, UV radiation field, and metallicity effects (Elmegreen 1993). Since $R_{\rm H}$ depends on the density of dust grains, H I envelopes are larger if the metallicity is lower. Since $G_{\rm surf}$ depends on the ambient UV radiation field, HI envelopes are larger near sites of active massive star formation. The pressure and radiation field, and consequently the molecular fraction, are expected to vary across spiral arms. In some spiral arms of M51 and perhaps M83, the H I peak is downstream from the CO peak, and spatially coincident with H II regions, consistent with a photodissociation origin for some of the H I (Allen, Atherton, & Tilanus 1986; Vogel, Kulkarni, & Scoville 1988).

4. Spiral structure in the outer disks

Since stellar disks form from gas disks which are largely molecular, much about the formation and evolution of disk galaxies could be understood by answering the following 2 questions. What determines the cold gas distribution? What physical conditions in the gas determine the rate at which stars form? Observational results reviewed here in §4–6 are beginning to provide some answers to these questions. The degree of dissipation experienced by molecular gas and its self-gravity strongly affect its behavior, yet are not well enough understood. Throughout galaxies, much of the gas behavior can be understood in terms of deviations from azimuthal symmetry in the gravitational potential, which perturb gas orbits, and consequently drive radial flows, cause gas pile-ups, and trigger star formation.

4.1. M51

The grand design spiral galaxy M51 has the strongest spiral density wave among nearby, well-studied galaxies, due to its ongoing tidal interaction with NGC 5195. Molecular line astronomers just cannot leave M51 alone. That is just as well since it continues to provide new insights into the evolution of the ISM as it passes through a strong spiral arm.

The CO spiral arms in M51 are $\simeq$1 kpc wide ($\simeq$20–30″), as judged from maps which combine interferometer and single dish data (Adler *et al.* 1992), and deconvolved single dish maps (Garcia-Burillo, Guelin, & Cernicharo 1993a). The average arm-interarm intensity contrast in CO of 3–6 (Adler *et al.* 1992; Garcia-Burillo *et al.* 1993a; Garcia-Burillo *et al.* 1993b) is about twice the arm-interarm contrast in the stars, as judged from 2μm maps (Rix & Rieke 1993). A factor of 2 difference in arm-interarm contrasts is not very much because random motions of the stars, whose orbits can cross, reduce the amplitude of the stellar arm below that of a gas arm (Binney & Tremaine 1987). Consequently the CO arm-interarm contrast can be understood via orbit crowding with only modest dissipation. There is no need for strong shocks or molecular cloud formation from H I .

Spiral arm streaming motions reach of a maximum of $\pm$100 km s^{-1} (in the plane of the galaxy), and are largely consistent with the density wave theory (Garcia-Burillo *et al.* 1993b; Rand 1993b). From high resolution (1–2″ = 50–100pc) CO and Hα observations, Rand (1993b) finds an asymmetric velocity profile across one arm, suggesting that cloud dissipation does occur.

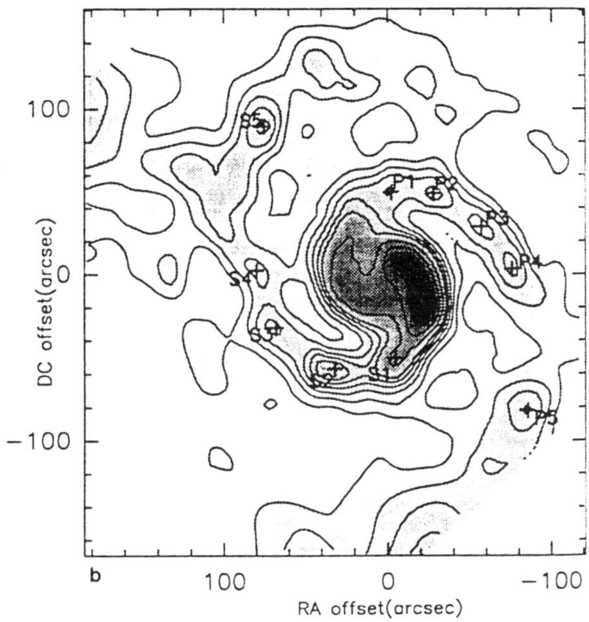

Figure 2. CO (2-1) intensity map of M51 at 12″ resolution, from IRAM 30-meter telescope. Notice the remarkable substructure along the molecular arms. Peaks are regularly spaced along each spiral arm, and have symmetrically located counterparts in the other arm. These are complexes of molecular gas which are probably located at resonances. From Garcia-Burillo *et al.* (1993a).

The existence of well-defined dust lanes which are spatially coincident with CO peaks and located upstream of the potential minimum in M51 (Vogel, Kulkarni, & Scoville 1988), suggests that some form of dissipation is occuring, although it seems to be a much weaker perturbation than a purely hydrodynamic shock.

The CO arms in M51 exhibit a number of discrete emission peaks spaced by 1.5–2 kpc, and symmetrically located with respect to the center, as shown in Figure 2 (Garcia-Burillo *et al.* 1993a). These are complexes of molecular gas, H I , and H II regions which are likely located at resonant positions. A strongly resonant gas response is one of the consequences of the enormous strength of the tidally-triggered density waves in M51. Rand (1993b) suggests that the 4:1 resonance may be responsible for an unusually large radial velocity shift in the western spiral arm, close to the location of one the resonant CO peaks identified by Garcia-Burillo *et al.* (1993a).

Line intensity ratios reveal differences in arm and interarm cloud properties (Garcia-Burillo, Guelin, & Cernicharo 1993a). Although the ^{12}CO(2-1)/(1-0) ratio is a constant 0.7–0.8 throughout the disk, the more optically thin ^{13}CO(2-1)/(1-0) ratio is 0.8 in the arms, 0.4 in the interarms, and 1.2 in the nucleus. The line intensity ratios are used to derive core-halo models of molecular cloud complexes, using LVG and Monte Carlo radiative transfer calculations. Cloud models derived for arm, interarm, and nuclear clouds suggest that the mass fraction in cores is <20% in interarms, 50% in arms, and 60% in the nucleus. The low ^{13}CO(2-1)/(1-0) ratio in the interarms indicates subthermal excitation and implies that there is a high fraction of low density diffuse gas in interarm regions.

Molecular column densities derived from the radiative transfer calculations of Garcia-Burillo *et al.* (1993a) are 3–5 times smaller than those from the standard CO–H_2 conversion factor – 3 times smaller for the arms, the nucleus, and the galaxy as a whole, and 5 times smaller for the interarm regions. This would suggest that the arm-interarm contrast in H_2 surface density is $\sim$2 times higher than that in CO emissivity. Comparisons by Adler *et al.* (1992) and Rand (1993a) of CO luminosities and virial masses for clouds also suggests that the standard CO–H_2 conversion factor overestimates H_2 masses by a factor of 2–3 in the arms of M51. While each of these techniques for estimating cloud masses is probably uncertain by a factor of 2–3, it is striking that 2 different techniques and 3 different groups get the same answer.

Star formation is triggered in the spiral arm density waves, although the precise mechanism by which this happens remains unclear. The Hα/CO ratio is $\sim$2 times higher on the arms than in the interarms (Lord & Young 1990; Garcia-Burillo *et al.* 1993a). Considering that dust extinction is probably higher in the arms, the star formation efficiency is enhanced in the arms by a factor of $\geq$2. The Hα and thermal radio continuum peaks are located downstream from the CO peak, and since the offset is seen in the thermal radio continuum (Tilanus & Allen 1988) as well as in Hα (Vogel, Kulkarni, & Scoville 1988), the offset is not merely the effect of extinction. The spatial offset has been interpreted as a time delay between cloud triggering at the site of maximum gas compression, and the emergence of young stars (Vogel, Kulkarni, & Scoville 1988; Tilanus & Allen 1988). Yet if this were true, the spatial offset should correlate with Ω-Ω_p, the difference between the gas angular speed Ω and the pattern speed Ω_p, and this is not clearly observed. Star formation may instead be initiated quickly throughout the density wave. H I and Hα are located together downsteam from the CO peak, suggesting that some of the H I is a photodissociation product (Allen, Atherton, & Tilanus 1986; Vogel, Kulkarni, & Scoville 1988; Rand & Kulkarni 1990).

What happens to dense gas as it enters and then leaves the spiral arms? Rand (1993a, 1993b) discusses possible formation and destruction mechanisms of the Giant Molecular Associations (GMAs) detected in CO interferometer maps of M51, with characteristic masses $\sim 3\times 10^7\ M_\odot$. The GMAs in the arms seem to be barely gravitationally bound, and probably become unbound once they leave the arms and enter a region of diverging streamlines and high shear. Due to orbital dynamics alone, spiral arms are regions of reduced shear and increased gas surface density (Roberts & Stewart 1987). Since the azimuthally averaged value of the Toomre stability parameter Q is close to 1 for the gas disks of spiral galaxies (Kennicutt 1989; chapter 7, p. 180), gas in spiral arms is particularly susceptible to gravitational collapse, and large self-gravitating gas complexes may be expected to form. Since the molecular gas entering the arm is presumably already lumpy and composed partly of self-gravitating GMCs with masses $\lesssim 10^6\ M_\odot$, the growth of even larger gas complexes in the arm can be considered a process of collisional agglomeration, although it is orbital dynamics which assembles large quantities of gas together in an environment where cloud collisions will be sticky.

4.2. M83

M83 has grand design spiral structure, although weaker density waves than M51. While the CO peaks are largely coincident with dust lanes in the spiral arms of M51, they are downstream from the dust lanes in at least one spiral arm segment of M83 (Wiklind *et al.* 1990; Lord & Kenney 1991). The difference is illustrated in Figure 3, which shows CO maps superposed on blue optical photos of the inner eastern arm of M83 and the inner western arm of M51. In this eastern arm of M83, the CO peaks are located together with Hα and H I peaks, roughly near the potential minimum of the arm. The difference between these arms may be related to the fact that M83 has smaller streaming motions and a weaker global density wave than M51. One possible explanation, due to Elmegreen (1988), is that in the weaker arm galaxy M83, the lower density part of the ISM is shocked as it begins to fall into the arm, while dense molecular clouds punch right through the shocks on nearly ballistic orbits. In M51, with its much stronger density wave, the shock is too strong for the molecular clouds to punch through, and all the gas piles up in a broad collision front.

Alternatively, Deutsch & Allen (1993) propose that the main difference between the arm segments in M51 and M83 is the CO–H_2 conversion factor. In both arm segments, the CO and non-thermal radio continuum peaks are located together, suggesting that cloud heating may be responsible for the difference. This explanation is suggested by the correlation between non-

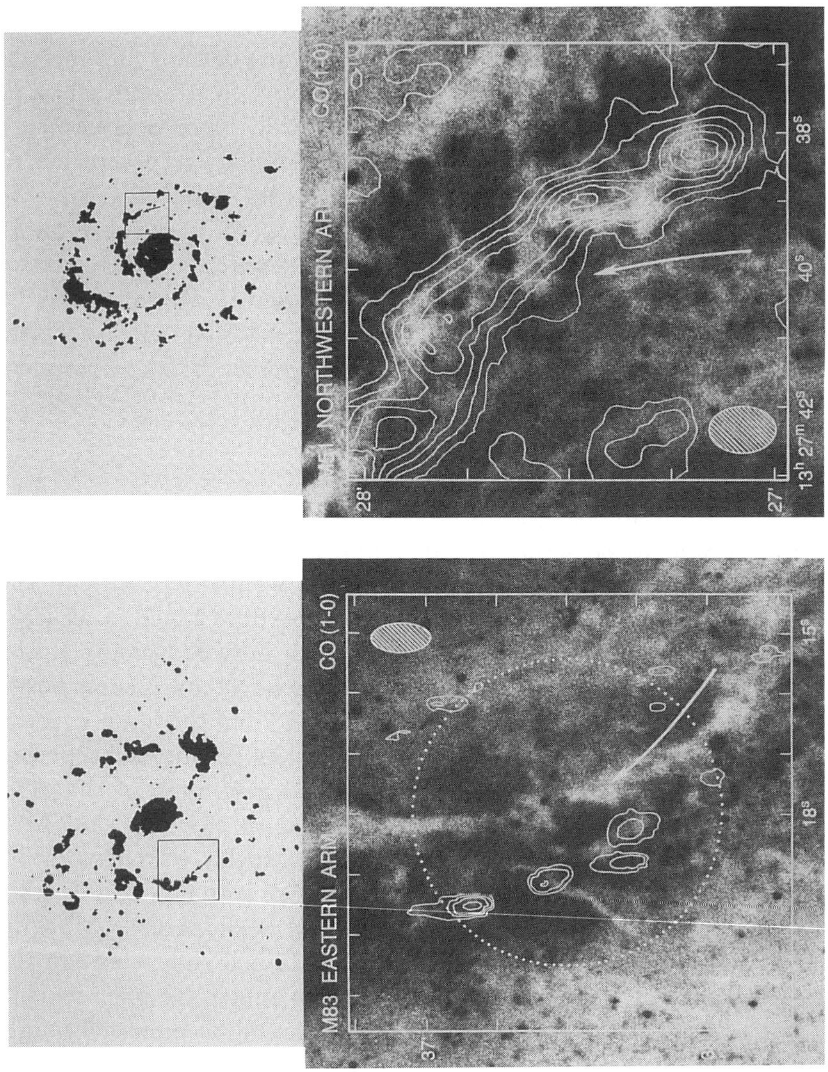

Figure 3. Interferometer CO (1-0) maps on blue optical photographs of spiral arm segments in M83 (left) and M51 (right). The CO peaks are located downstream from the dust lane in the M83 arm, and coincident with the dust lane in the M51 arm. The arrows indicate the direction of circular orbits. From Lord & Kenney (1991).

thermal radio continuum and CO emission in galaxies (Adler, Allen, & Lo 1991; Garcia-Burillo *et al.* 1993a), which may indicate that cosmic rays both produce the synchrotron radiation (high energy electrons) and heat the molecules in the centers of cold opaque colds (low energy protons and nuclei). M83 has a higher supernova rate than M51 (6 vs. 0 recorded since 1923), making it plausible that recent supernovae create the strong non-thermal radio continuum emission observed near the H II regions and CO peaks in M83. In M51, the radio continuum emission near the H II regions is mostly thermal, and the strongest non-thermal emission is on the dust lanes. Yet even if the difference in CO spiral structure is partly a heating effect, there are still physical differences between the arms which need to be explained. Why is the non-thermal radio continuum strong in the M51 dust lane but weak in M83's dust lane? Why is the massive star formation rate in the M83 arms several times higher than the M51 arms? Why is the dust lane in the eastern arm of M83 smoother than its counterparts in M51 (Lord & Kenney 1991)?

The eastern arm region of M83 which has been the focus of many studies may not be characteristic of all spiral structure in M83. Closer to the nucleus along the same arm in M83, the strongest non-thermal radio continuum emission coincides with dust lanes, but in the "eastern arm", it coincides with the ridge of H II regions and CO, and is downstream from the dust lane (Deutsch & Allen 1993). Indeed, the grand design spiral galaxy M100 exhibits both types of behavior. In some spiral arm segments, CO is coincident with the dust lanes, and in others the CO peak is downstream from the dust lane and coincident with H II regions (Rand 1995). Is this difference due to the speed at which gas enters the spiral arm? Clearly there is more to learn about the behavior of the multi-phase interstellar medium passing through a density wave.

4.3. M101

The CO morphology in the multi-arm Sc galaxy M101 provides evidence of how the dense ISM in spiral galaxies is organized on 100–1000 pc scales in the absence of strong density wave forcing. Figure 4 shows that M101 has a central molecular bar surrounded by a partial ring and several molecular spiral arms or arm fragments (Kenney *et al.* 1997). One or more of these arms may be weak density waves, although most of them are irregular and fragmentary. Virtually all of the CO peaks are associated with dust lanes, and the CO apparently traces the principal mass concentrations within more extended spiral-shaped filaments of gas. A comparison with an Hα map shows that some of the weaker CO arm fragments have little or no associated massive star formation. The molecular arms may grow by

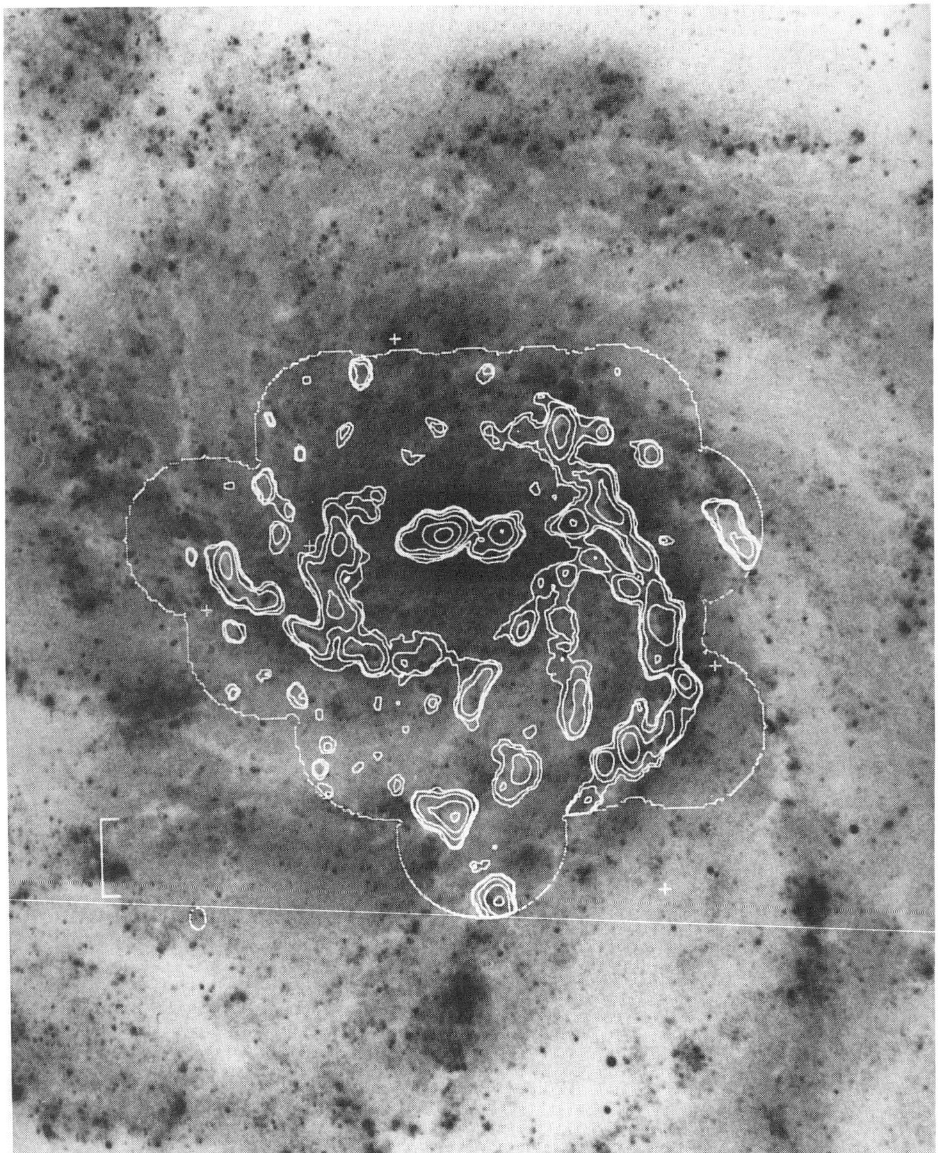

Figure 4. CO (1-0) interferometer mosaic map at 9″×7″ resolution of central 6 kpc of M101, on blue optical image. A bar, a ring, and numerous spiral features are evident in the molecular gas distribution. Virtually all of the CO features coincide with dust lanes. From Kenney *et al.* (1997).

local instabilities or sticky collisions and get stretched by shear into spiral shapes. Even the smallest resolved features, with sizes of ∼200 pc, and gas masses $\sim 10^6\ M_\odot$, are elongated, show spiral structure, and are apparently influenced by differential rotation. In both M101 (Fig 4) and NGC 1068 (Fig 12, Planesas *et al.* 1991), molecular gas rings which appear to be composed of tightly wound spiral arm segments are located at the ends of stellar bars.

5. The central few kiloparsecs

5.1. WEAKLY BARRED GALAXIES

Bar-like molecular gas morphologies are found in the central 1–few kpc of many weakly barred galaxies. Although they are sometimes referred to casually as "molecular bars", there may be 2 physically distinct situations in which highly elongated central gas concentrations are observed. In some cases, the gas clearly lies along the leading (downstream) edges of the stellar bar, as illustrated in Figure 5 for the Sc galaxy M101 (Kenney *et al.* 1997). In the case of the Sc galaxy IC342, the stellar distribution appears axisymmetric, and the slightly curved CO features have been interpreted as loosely wound gaseous spiral density waves (Turner & Hurt 1992). This is plausible because of the high gas mass fraction, which increases from ∼10% at at a radius of 500 pc to ∼50% at at a radius of 100 pc. Observationally it is not always easy to distinguish between the 2 situations. Gas responds strongly to even weak stellar bars, and weak stellar bars may not always be obvious in the presence of dust and star formation, even in the near-infrared (Regan & Vogel 1995). The curvature of the gas concentrations does not by itself suggest a spiral arm, since gas concentrations can also be curved as a response to a weak stellar bar (Athanassoula 1992).

The z-shaped CO morphology in the central 1 kpc of M101, which is highlighted in Figure 5a (Kenney *et al.* 1997), is similar to the gas distribution observed in other weakly barred Sc galaxies. Figure 5c shows that the largest CO concentration lies along the leading edge of a weak stellar bar, as revealed by R-band isophotes. The gas has responded strongly to a small deviation from azimuthal symmetry in the stellar potential. The gas distribution is similar to both hydrodynamic simulations (Roberts, Huntley & van Albada 1979; Athanassoula 1992) and N-body simulations with sticky cloud-cloud collisions (Combes & Gerin 1985). Most of the molecular gas in the the central kpc of M101 is located in the central $r = 15''$, where the stellar distribution is nearly axisymmetric. There is little gas at $r = 15'' - 30''$, where the potential is maximally elongated. The gravitational torques exerted on the gas by the stars are strongest here, and the gas would be driven inwards from this region.

The kinematics of the gas along the bar in M101 are revealed by the

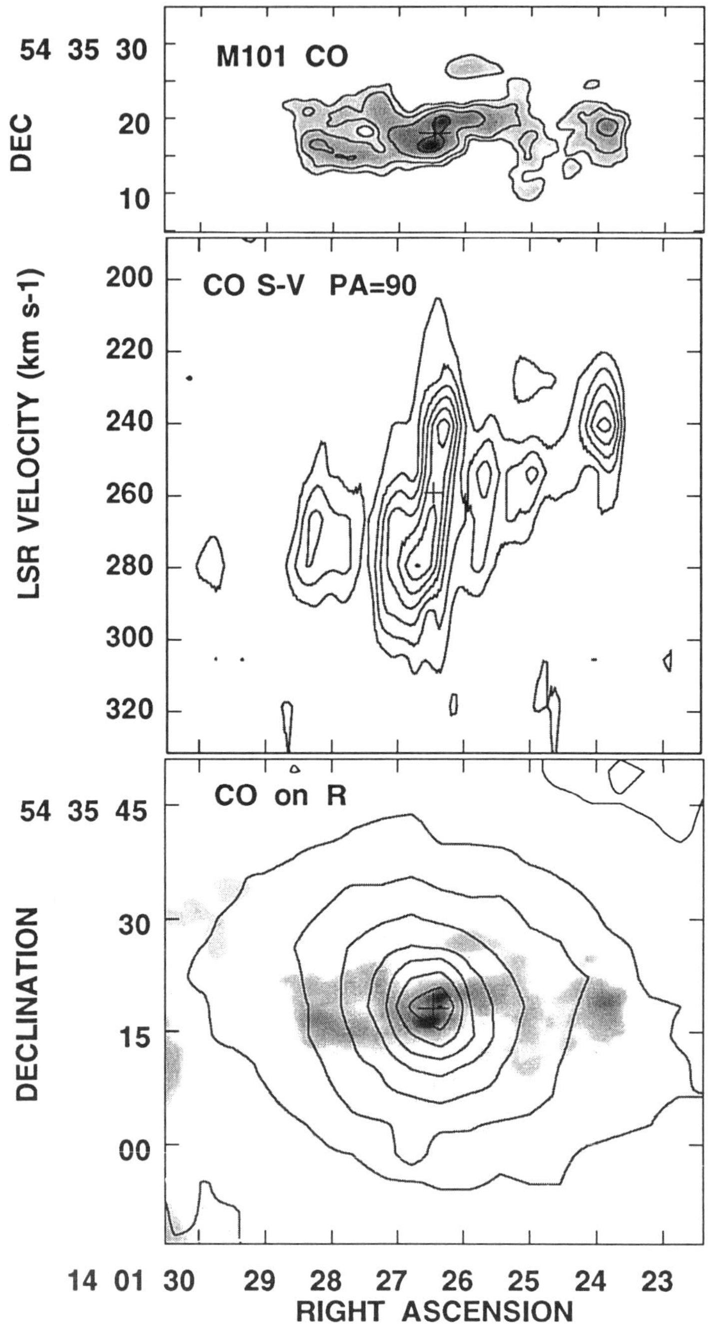

Figure 5. a.) CO (1-0) map of M101 at $3.0'' \times 2.5''$ resolution obtained with Owens Valley interferometer (Kenney *et al.* 1997). b.) CO spatial-velocity map along PA = 90° , which is 45° from the kinematic major axis of the outer disk. Note abrupt change in gas kinematics at $r \simeq 3'' \simeq 80$ pc. The region of greatest gas surface density corresponds to the outermost part of the high velocity component. c.) CO grayscale map and R-band contour map in central $r = 30'' = 800$ pc of M101. The modest oval distortion in the stellar potential is presumably responsible for the molecular gas piling up in nearly linear ridges along the leading edge of the bar.

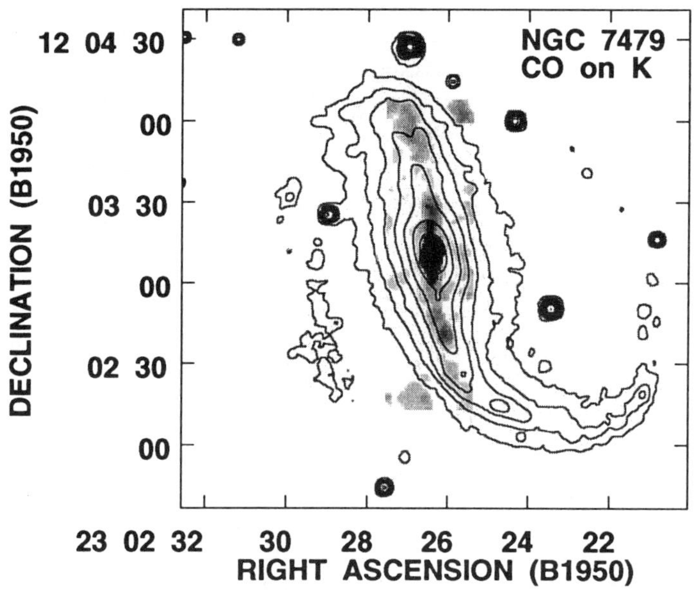

Figure 6. K-band contour map and grey-scale CO map of the barred Sc spiral NGC 7479, which has an unusually gas-rich bar. The molecular gas is offset from the center of the stellar bar, creating a torque which drives gas radially inward at a net speed of 10–20 km s^{-1}. Adapted from Quillen *et al.* (1995).

spatial-velocity diagram in Figure 5b, which is a cut along PA = 90°, nearly along the bar major axis (~80°), and 45 degrees from the line of nodes (PA = 45°). Two distinct kinematic components are apparent – a high-velocity circumnuclear component in the central $r = 4''$, and a lower velocity component at $r = 4-15''$, corresponding to the nearly linear ridges. Near the major axis of many galaxies, there is a high-velocity circumnuclear component with mostly circular motions surrounded by a region of lower gas velocities with larger non-circular motions (Handa *et al.* 1990; Koribalski, Dickey, & Mebold 1993). The abrupt transition from high to low velocity gas marks a convenient outer boundary to the circumnuclear region. This kinematic transition generally occurs at or near inner Lindblad resonances, which in many galaxies is the region of maximum gas surface density (see §6.1).

5.2. STRONGLY BARRED GALAXIES AND GAS INFLOW ALONG BARS

Gas driven radially inward by stellar bars causes circumnuclear starbursts, builds bulges, and redistributes gas enough to nearly eliminate chemical abundance gradients (Martin & Roy 1995). Although large non-circular motions are predicted and observed along bars, it is non-trivial to determine the net inflow rate of gas from the observed local streaming motions. Gas has the largest inward motion where it piles up and is easiest to detect, but is predicted to move outwards on other parts of its orbit, where the gas surface density is much lower (Athanassoula 1992). One way to estimate the net inflow rate is to calculate the gravitational torque exerted on the gas by the stars, by computing the spatial offset of the gas from the stars. Quillen *et al.* (1995) have done this for the barred galaxy NGC 7479, which has an unusually gas-rich bar. Figure 6 shows a 2μm image of NGC 7479 superposed on a CO map. The molecular gas is offset from the center of the stellar bar by a typical distance of a few hundred pc, creating a torque which drives gas radially inward at a net speed of 15 km s^{-1}, and a net rate of 4 $M_\odot$ yr^{-1}. This net inflow speed is much less than the maximum local streaming motions of ~200 km s^{-1}, but still fast enough that the gas which is now at the bar end will reach the circumnuclear region in only 3 orbits, or 6×10^8 yr, tripling the mass within the central 200 pc. This inflow speed is faster than most of the quasi-steady-state bar models of Athanassoula (1992), in which net inflow speeds are only ~1–2% of the maximum local radial streaming motions. These models probably correspond with older and more evolved bars. The inflow speeds in NGC 7479 are similar to those in minor merger models (Mihos & Hernquist 1994; Hernquist & Mihos 1995). Since NGC 7479 has highly asymmetric spiral structure, an off-center nucleus, a gas-rich bar, and a circumnuclear starburst, it could well be a minor merger with a young bar.

Large torques exerted by strong stellar bars drive gas inward rapidly, so that gas will not remain long in a strongly barred region. Thus old strong bars are not expected to have much gas, while young strong bars or weak bars may. Sc's (§5.1) tend to have more gas in the region of the bar than Sb's and Sa's (§6.1), because bars are weak in most Sc's (Elmegreen & Elmegreen 1985; Combes & Elmegreen 1993), so the torques are smaller and the gas is not driven radially inward as rapidly. Some stars do form along the bar, particularly in young bars (Martin & Roy 1995; Friedli & Benz 1995), and these stars are injected into the galaxy with the large non-circular streaming motions of the gas. The stars born on these elongated orbits may help build the stellar bar.

5.3. BAR ENDS

Prodigious star formation occurs in many galaxies at the sites where bars end and spiral arms begin. The morphology and kinematics of a large molecular complex at a bar end in M83 provides insight into the gas dynamical processes which cause this enhanced star formation activity (Kenney & Lord 1991). The cloud complex is resolved into two main components, separated by 250 pc, and aligned radially within the galaxy. The inner one is associated with the bar and the outer with the spiral arm. The tangential velocity of the bar component is 60 km s^{-1} (in the plane of the galaxy) less than that of the spiral arm component, indicating large tangential streaming motions in one or both. It is likely that there are also large radial motions which form the complex by bringing the 2 gas streams together. Outward moving gas from the bar region crowds together with gas from the outer disk, increasing the gas density and triggering star formation. The increase in gas density is due to orbit crowding, since the gas is funneled through a small region as a result of perturbations from circular orbits caused by the non-axisymmetric mass distributions in the bar and spiral arm. A similar effect has been observed in M100 (Rand 1995).

Bar ends are where 2 streams of gas meet and one of them undergoes strong star formation. At the bar end in M83 virtually all of the star formation is occuring in spiral arm gas (Kenney & Lord 1991). The bar gas itself has a low star formation efficiency: the UV/CO ratio in M83 is several times lower along the bar than the circumnuclear region or the outer disk, even though the bar in M83 contains a lot of molecular gas (Handa *et al.* 1991). Possible reasons for a low star formation efficiency along bars are discussed by Elmegreen (1979), Tubbs (1982), and Kenney & Lord (1991). Of course, some galaxies do harbor strong star formation along the bar, including NGC 7479 (Quillen *et al.* 1995) and NGC 3359 (Martin & Roy 1995). It will be interesting to compare the gas kinematics along those bars with and without strong star formation, to see whether the difference can be understood in terms of instabilities or cloud-cloud collisions.

6. Circumnuclear regions

6.1. CIRCUMNUCLEAR RINGS AND INNER LINDBLAD RESONANCES

Many spiral galaxies have a compact circumnuclear gas component, which often lies deep within a bar and coincident with the central part of a traditional stellar bulge. These circumnuclear gas disks are generally distinct from surrounding gas, both spatially and kinematically, and often contain a significant fraction of the enclosed dynamical mass. The spatial relationships between the circumnuclear CO and the stellar morphologies are illustrated in Figures 7 and 11 for NGC 3351 and NGC 891.

In the nearly face-on, strongly barred Sb galaxy NGC 3351 (Kenney *et al.* 1992), the largest gas concentration is located at a radius of 340 pc, approximately ~10% of the bar radius, and ~50% of the radius at which the spheroidal bulge begins to dominate the stellar light profile (see Figure 7). The strongest CO emission arises where inward flowing gas from the bar region, traced by dust lanes and faint CO emission and along the leading

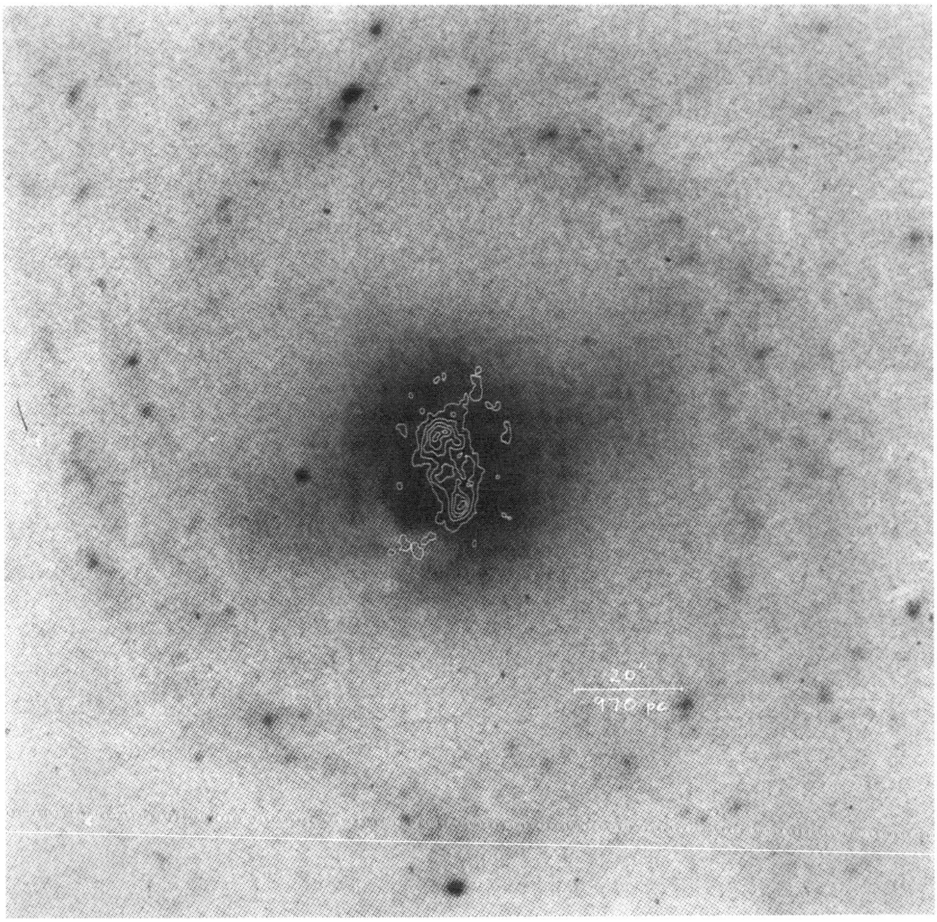

Figure 7. CO contour map on blue optical image of NGC 3351. The strongest CO emission arises from 2 peaks oriented perpendicular to the large-scale bar, located close to an inner Lindblad resonance, and within the bulge-dominated region. There is not much molecular gas along the bar, although weak CO emission coincident with dust lanes can be traced along the leading edge of the stellar bar. From Kenney *et al.* (1992).

edge of the stellar bar, merges with gas on more circular orbits at the inner Lindblad resonance (ILR).

An edge-on view of a circumnuclear disk is shown in Figure 11 for the possibly barred Sb galaxy NGC 891 (Scoville *et al.* 1993). The circumnuclear component of NGC 891 contains 3×10^8 $M_\odot$ of H_2 within a radius of 225 pc, and is well-defined in the spatial-velocity diagram shown in the bottom panel of Figure 11. Gas velocities rise steeply, with nearly solid body rotation, to a maximum line-of-sight velocity of ±200 km s^{-1} at $r = 5'' = 225$ pc from the nucleus. There is very little gas with line-of-sight velocities of ±200 km s^{-1} just beyond the circumnuclear disk, principally due to a lack of gas between 300 pc and 3 kpc. This circumnuclear gas disk is deep within a boxy bulge, which may be an edge-on bar similar to the one in the Milky Way, which has thickened due to an instability (Raha *et al.* 1991) or vertical scattering near resonances (Combes *et al.* 1990).

There is a great variety in circumnuclear gas morphologies, as illustrated by the CO maps in Figure 8 for seven galaxies with bars of at least moderate strength. There are rings or partial rings (NGC 4314), filled exponential disks (NGC 3504, spiral arms (NGC 6951), "twin peaks" (NGC 3351), NGC 4102), or some combination of these. Sc galaxies, particularly those with weaker stellar bars, have strong CO emission along the linear or slightly curved ridges which lead into the circumnuclear region (M101, M83), while the Sa (NGC 4314) and Sb (NGC 3351, NGC 3504, NGC 4102, NGC 6951) galaxies have weak CO emission along the bar. This variety is only partially understood, although the largest gas concentrations are often located near inner Lindblad resonances (ILR). The large gas concentrations near ILRs can be understood as a combination of orbit crowding and orbit population effects. Gas orbits are tightly crowded near the ILR, since the periodic stellar orbits change their orientation by 90° here. Since radial inflow stalls near the ILR, there is a lot of gas populating these orbits.

The radial inflow rate of gas, which is large along bars, is predicted to be much lower near ILRs (Combes & Gerin 1985; Combes 1988; Shlosman, Frank, & Begelman 1989), thereby creating gas concentrations located approximately between the Outer Inner Lindblad Resonance (OILR) and the Inner Inner Lindblad Resonance (IILR). (The position of the two ILRs are not so precise unless the potential is very nearly axisymmetric. If the potential is highly non-axisymmetric, a single orbit can have its small axis at the IILR and the great axis at the OILR, and it is only meaningful to speak of a nuclear ring in the ILR region. See Athanassoula (1992) for discussion.) The gravitational torques exerted by the stars on the gas, which drive radial gas flows, are much smaller between the OILR and IILR, or may even act to drive gas outward toward the OILR, depending on the shape of the potential and other poorly understood details. Rotation curves

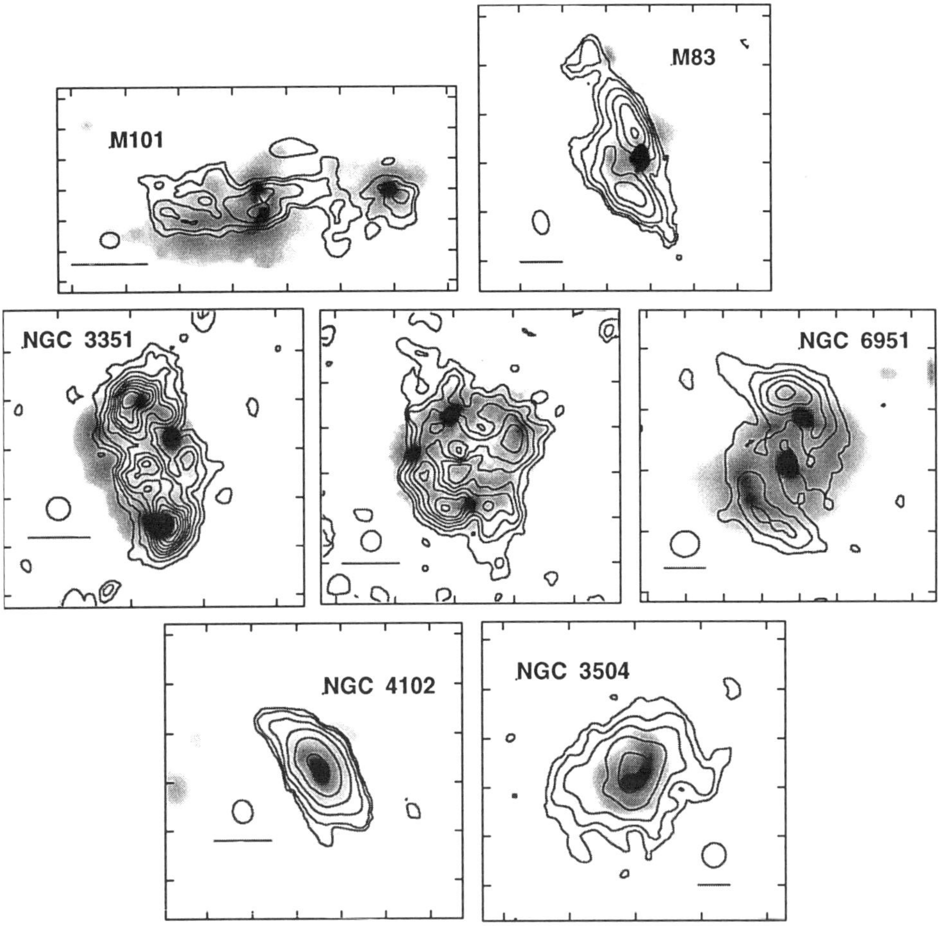

Figure 8. CO contour and Hα+[N II] greyscale maps of the circumnuclear regions of 7 galaxies, which illustrate the variety of circumnuclear gas morphologies. All galaxies have large-scale stellar bars which extend beyond the areas shown in these maps. The strongest CO emission is located close to Inner Lindblad Resonances in NGC 3351, NGC 6951, NGC 4314, and perhaps also in M101 and M83. In the young starbursts NGC 3504 and NGC 4102 the gas is centrally concentrated, well inside OILRs and perhaps inside IILRs. NGC 3351, NGC 6951, M101, and NGC 3504 from Kenney *et al.* (1992), NGC 4314 from Benedict, Smith & Kenney (1996), NGC 4102 from Jogee & Kenney (1995, 1997). Beamsizes, which are typically 2–3″ (except 7″ × 4″ for M83), are shown as ellipses. Vertical tickmarks are separated by 5″ (except 15″ for M83). The horizontal line in each panel is 300 pc in length.

derived from gas kinematics (Arsenault *et al.* 1988; Kenney *et al.* 1992) or azimuthally averaged stellar light profiles (Combes *et al.* 1992) confirm that many of these gas concentrations are indeed located near ILRs. In M100 (Knapen *et al.* 1995; Rand 1995; Sakamoto *et al.* 1995) and NGC 4314 (Benedict *et al.* 1993), small scale nuclear bars which are aligned with the large scale stellar bars are tucked inside the gas rings, which is expected inside the IILR (Binney & Tremaine 1987), and confirms the approximate ILR locations predicted from gas kinematics. The gas kinematics can change abruptly at the ILR, as shown in the spatial-velocity plot of M101 in Figure 5b. Although there are large non-circular gas motions in the bar regions of galaxies, the motions are often observed to be more circular between the ILRs. An excellent example of this is NGC 4314. Within the CO ring shown in Fig. 8, the kinematics are predominantly circular. Yet in the dust lane component which leads into the ring and lies along the minor axis, there are inward radial motions of 20–70 km s^{-1} (in the plane of the galaxy) (Benedict, Smith & Kenney 1996). This is clear evidence of inflowing gas feeding an ILR ring.

Molecular spiral arms which are located near ILRs exist in M100 (Rand 1995; Sakamoto *et al.* 1995), NGC 6951 (Kenney *et al.* 1992), and NGC 1530 (Regan, Vogel, & Teuben 1995). Hydrodynamic simulations by Athanassoula (1992) produced circumnuclear spiral features near ILRs, and from simple dynamical considerations, Combes (1988) and Wada (1994) described why spiral gas patterns are sometimes expected near ILRs. Closed stellar orbits change orientation abruptly by 90° at every principal resonance. Due to dissipation, gas orbits change more gradually between resonances, forming spiral gas concentrations similar to kinematic density waves. If there are 2 ILRs, the axes of the gas orbits change with radius in such a way as to make trailing spiral arms near the OILR and leading spiral arms near the IILR (Wada 1994). The CO arms in M100, NGC 6951, and NGC 1530 are all trailing, and are presumably located near OILRs. The circumnuclear arms in M100 exhibit non-circular streaming motions similar to those in outer disk spiral arms (Sakamoto *et al.* 1995). Leading gaseous arms near the IILR have not yet been observed, although near-infrared images of M100 suggest modest leading arms in the stellar distribution in the vicinity of the putative IILR (Knapen *et al.* 1995). The CO may not be distributed in a spiral even if some other component of the ISM is. There is a trailing nuclear spiral in the B–I color map of NGC 3504, presumably due to dust, although the CO morphology is nearly azimuthally symmetric and shows little evidence for a spiral (Kenney, Carlstrom, & Young 1993).

In NGC 3351, the strongest CO emission arises from "twin peaks", located symmetrically about the nucleus, and oriented perpendicular to the large-scale bar (Kenney *et al.* 1992). These peaks are where inward flowing gas from the bar, traced by faint CO emission coincident with the dust lanes along the leading edge of the stellar bar, appears to spatially intersect a ring of gas on more circular orbits at the ILR (see Figures 7, 8). The twin peaks are probably similar to bar ends, since they are both locations where 2 kinematically distinct gas streams meet (see §5.3). This phenomenon appears distinct from spiral arm formation. Although the twin peaks are particularly prominent in NGC 3351, similar features can be recognized in Figure 8 in M101 and NGC 6951. The region of strongest CO emission in the northern CO spiral arm of NGC 6951 is where the CO spiral arm intersects a partial ring of CO and Hα emission. CO spectra near this northern peak show broad (FWHM$\geq$100 km s^{-1}), complex line profiles with 2 or more peaks, perhaps indicating the kinematics of 2 interacting gas streams. The strength of the "twin peaks" component may depend on how much gas is inflowing compared to how much gas is already in the ILR ring, as well as the following dynamical differences.

The variety in gas morphologies is partly due to variations in the gravitational potential, including the degree of central concentration and also the strength of the bar. The rotation curve and the pattern speed of the stellar bar determine the number, location and separation of the ILRs. Gas behavior should also depend on whether there are small scale nuclear bars, and if so, whether these rotate faster than the large scale bars (Shlosman *et al.* 1989; Wozniak *et al.* 1995), or are part of the same pattern (Knapen *et al.* 1995; Benedict *et al.* 1993). Gas viscosity and self-gravity are the critical ISM parameters. The effective viscosity of the gas influences how much the gas orbits differ from the closed stellar orbits (Wada 1994). If the gas comprises more than $\sim$10% of the dynamical mass, the gas self-gravity may help drive radial inflow (Wada & Habe 1992; Heller & Shlosman 1994). Gas mass fractions as high as 10–90% are estimated for the central few hundred parsecs of many galaxies (Kenney *et al.* 1992; Turner 1994; Scoville *et al.* 1994). Some of the variety is undoubtedly due to evolutionary effects. Timescales are much shorter in circumnuclear regions: starburst timescales are 10^7–10^8 yrs, and dynamical timescales are $\sim10^7$ yrs. Both the gas inflow rate and the gas consumption rate due to star formation probably vary in time. Of course, dynamical and evolutionary effects are coupled together. Gas inflow builds the central mass concentration, which steepens the rotation curve, eventually forming ILRs, whose locations continue to change as the central mass concentration grows. While a wealth of gas morphologies can be understood in principal, it has not yet been demonstrated which factors are most responsible for the existing variety.

Star formation appears to be triggered at the "twin peaks", perhaps in much the same way it happens at bar ends – by orbit crowding or the collision of 2 gas streams (see §5.3) – due ultimately in both cases to deviations from azimuthal symmetry in the large-scale potential of the galaxy. The short rotation period in circumnuclear regions implies large angular offsets between the locations of star formation triggers and the appearance of luminous H II complexes. A 10^6 yr time delay corresponds to 60° in azimuth, for a rotation speed of 200 km s^{-1} at a radius of 200 pc. Consequently, the relationship between the triggering site and the H II complex is not as clear as it is in outer disk spiral arms, where a 10^6 yr time delay typically corresponds to less than a few degrees in azimuth. Even if all the star formation were triggered at "twin peaks", the rapid drift of gas away from these sites could result in H II complexes peppered throughout a ring, as star formation propogates through a large gas complex. This might explain why NGC 3351 has a ring of "hot spots", including 2 near the CO "twin peaks". A ring of "hot spots" might also be explained by a gravitationally unstable gas ring (e.g., Elmegreen 1994).

What is the ultimate fate of gas which is driven radially inward along bars? Except when the bar is young (Martin & Roy 1995; Friedli & Benz 1995), present data suggests that very little gas is consumed by star formation along the bar, and much of it is consumed by star formation near the ILRs. In some galaxies, strong CO is located at larger radii than the rings of intense star formation (Ishizuki *et al.* 1990; Turner & Hurt 1992; Kenney *et al.* 1992; Rand 1995). Figure 8 illustrates that the regions of star formation are more centrally concentrated than the CO distributions (note especially NGC 6951 and M83). Apparently the inward flowing gas does not experience vigorous star formation until it reaches the vicinity of ILRs, where its inward flow slows, and gas orbits become more circular.

The nearest galaxy which shows evidence for this kind of behavior is the Milky Way. Figure 9 shows longitude-velocity distributions of CO(J = 2-1), H I and 2 tracers of star formation in the central ±3° (±450 pc) of the Milky Way, from Oka *et al.* (1995). The central 2° (projected $R = 300$ pc) harbors a massive molecular gas complex which exhibits non-circular motions. The dense gas has a highly asymmetric distribution about the nucleus, indicating that it is not in equilibrium, but may be flowing inward in lumps. Both H II regions and OH/IR stars, which trace star formation over timescales of 10^7 and 10^8 yrs, are highly concentrated to the central 100 pc. Both show solid body rotation curves with no evidence for non-circular motions, although the OH/IR stars have a much steeper velocity gradient and reach 200 km s^{-1} at $R \sim 100$ pc. Since these are older and more dynamically evolved, they are presumably a better tracer of the rotation curve than gas or H II regions. This central star forming disk, which has undergone star formation, intermittently or continuously, over the last $\geq 10^8$ yrs (Oka *et al.* 1995), is spatially coincident with the inner

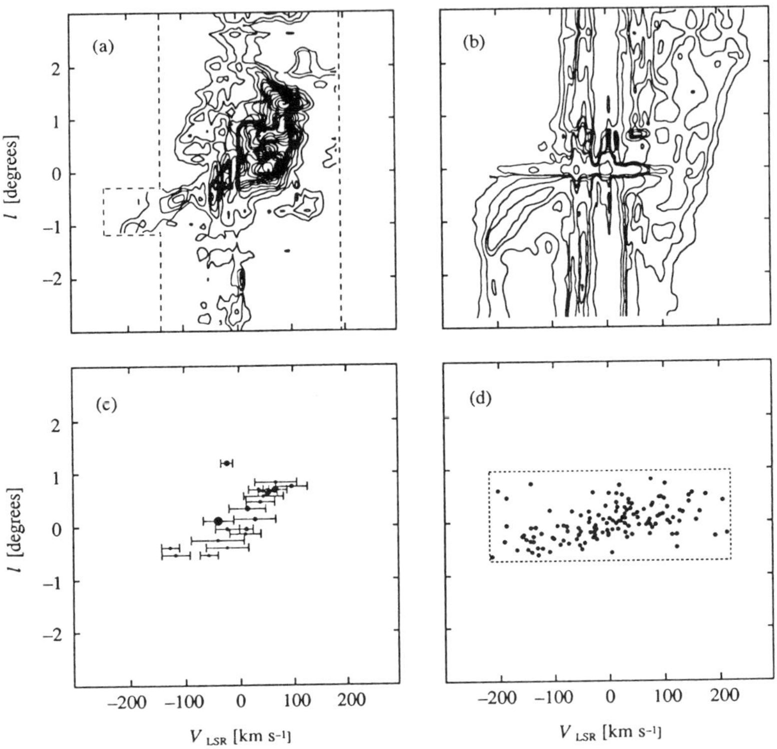

Figure 9. Longitude-velocity distributions of gas and tracers of star formation in the central $\pm 3^\circ$ (± 450 pc) of the Milky Way. a.) CO(J = 2-1) emission at b = 0° (Oka *et al.* 1995). b.) HI emission at b = 0° (Sanders *et al.* 1977). c.) HII regions traced by the H109α emission line (Pauls & Mezger 1975). d.) OH/IR stars traced by OH maser line (Lindqvist *et al.* 1991). Dashed rectangle shows l–V coverage of survey.

part of the dense gas complex. Most star formation does not occur until gas reaches this central disk, whose outer edge Oka *et al.* (1995) identify as an Inner Lindblad Resonance. The circumnuclear region of the Milky Way is therefore similar to other barred galaxies in terms of the gas kinematics and the pattern of star formation.

6.2. STARBURSTS

Those galaxies in the early stages of the most intense circumnuclear starbursts have their gas centrally concentrated. Figure 8 shows that the young starbursts NGC 3504 and NGC 4102 have CO distributions which are centrally peaked and nearly azimuthally symmetric, in contrast with the 5 "non-starbursting" galaxies in this Figure. (While the term "starburst" is used casually in the literature, we use "starburst" to mean those galaxies which are comparable to M82 in their circumnuclear luminosity (of some suitable tracer of star formation), and their ratio of luminosity to gas mass. Galaxies with significantly lower circumnuclear luminosities and luminosity to gas mass ratios are referred to here as "non-starbursts".) In these 2 starburst galaxies the CO radial distributions are nearly exponential with scale lengths of $\sim$300 pc (Kenney *et al.* 1993; Jogee & Kenney 1995, 1997). These starbursting molecular gas disks are an order of magnitude more compact than the exponential molecular gas disks which inhabit the outer disks of many Sc galaxies (Young & Scoville 1982).

In these young starbursts, most of the gas lies within the region where the rotation curve is rising steeply, and where the gas has nearly circular orbits. The gas depletion timescale, inferred from rate of star formation per unit gas mass, is shortest in the center, where the rotation curve rises steeply and is nearly solid body (Kenney *et al.* 1993). Consequently, starbursts evolve from the inside out. Intense star formation where the rotation curve is nearly solid body has also been observed in the center of Maffei 2 (Hurt & Turner 1991) and in disturbed galaxies (Keel 1993). The gas surface density (assuming the 'standard' CO–H_2 value) is approximately equal to the Toomre critical surface density in the starburst region of NGC 3504 (Kenney *et al.* 1993), so the simple gravitational instability theory of star formation seems to apply in circumnuclear starbursts as well as outer disks (Kennicutt 1989; chapter 7).

Are these young starbursts dynamically different from "non-starbursts" which have rings of gas and star formation between ILRs? NGC 3504 seems to have ILRs, as judged both from its rotation curve and its dust lane morphology. The region containing most of the gas and star formation is inside the OILR, although it is not clear whether it lies between the OILR and IILR or extends inside the IILR (Kenney, Carlstrom, & Young 1993). It may be that the largest gas concentrations and the most intense star formation are generally confined to the annulus between ILRs (Arsenault *et al.* 1988; Kenney *et al.* 1992; Knapen *et al.* 1995), and the region between the ILRs is wide in some galaxies. Alternatively, perhaps the most intense starbursts occur only when large quantities of gas get inside the IILR.

The behavior of gas inside the ILR is poorly understood, yet is critical for understanding starbursts. Although gravitational torques are the dominant radial driver of gas along large-scale stellar bars (Combes 1988), other effects might become dominant inside the ILR. If rapidly rotating, small-scale nuclear bars form inside the ILR, gas may be driven further inward by gravitational torques (Shlosman, Frank, & Begelman 1989). Gas might collapse rapidly inside the IILR once the gas mass is sufficiently large to become self-gravitating (Wada & Habe 1992; Heller & Shlosman 1994). Gas may drift toward the center as a consequence of dynamical friction on Giant

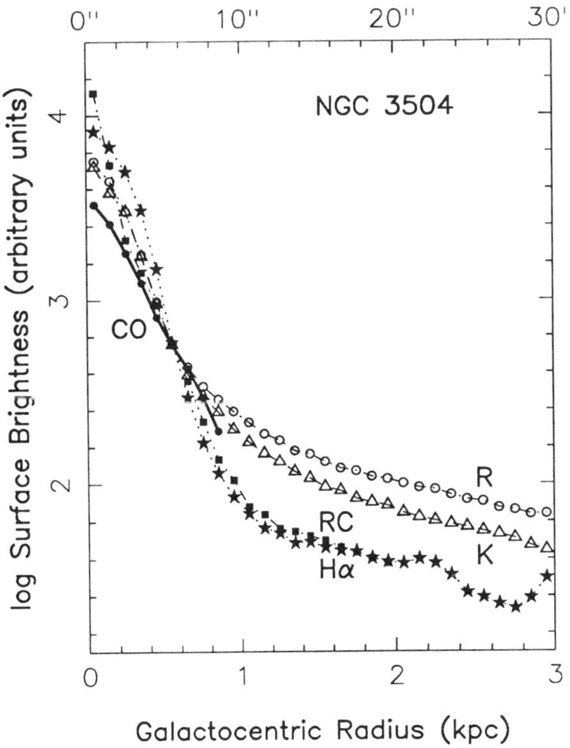

Figure 10. Radial distributions of CO, Hα+[NII], K-band, R-band, and 1.49 GHz radio continuum (RC) emission in the central 3 kpc of the starburst galaxy NGC 3504. The vertical scale gives the H_2+He surface density in $M_\odot$ pc^{-2}, for the standard CO–H_2 conversion, and other profiles are normalized to agree at r = 5.5″. The CO profile is well fit by an exponential with a scale length of 300 pc. This compact starbursting molecular gas disk should form a compact stellar disk. From Kenney, Carlstrom, & Young (1993).

molecular clouds (Binney & Tremaine 1987; Stark *et al.* 1991; Combes *et al.* 1992) or viscous torques (Lesch *et al.* 1990). Alternatively, perhaps large quantities of gas are driven deeply toward the nucleus only when there is a large dynamical disturbance, such as a tidal interaction or merger, and orbits are so perturbed and transient that there is no meaningful ILR. An ILR might form after the gas has been driven in.

The starbursts in NGC 3504 and NGC 4102 appear to be in very early stages of evolution. Their CO distributions are not ringlike despite the fact that gas is being consumed more rapidly in the center (Kenney, Carlstrom, & Young 1993; Jogee & Kenney 1995, 1997). NGC 4102 has a compact ionized gas outflow with expansion timescale of only 10^6 yr (Boer 1994). In these galaxies, the starburst has not yet consumed or blown out much of the central gas. Since starbursts consume their gas most rapidly in the center, and also generate outflowing winds, they are predicted to have ring-like gas distributions at later times. NGC 2782, which has an outflow timescale of 4×10^6 yr (Boer 1994), has a ringlike CO morphology (Ishizuki 1994). The well-studied, nearby, edge-on galaxies M82 and NGC 253 are also evolved starbursts, containing freely expanding winds with outflow timescales of 3–10×10^6 yr (Heckman, Armus, & Miley 1990). M82 has a two-lobe (edge-on torus?) structure in CO, as well as the dense gas tracers HCN and HCO^+ (Carlstrom 1988). However, it is more centrally peaked in optically thin $C^{18}O$ and 450μm continuum emission, which traces cold dust (Smith *et al.* 1990; Wild *et al.* 1992). Since different tracers give different answers, this places some uncertainty on the detailed molecular gas distribution in M82. Although M82 and NGC 253 have similar IR and CO luminosities, many other molecules (CS, SiO, HNCO, CH_3OH, CH_3CN, HCN) are at least an order of magnitude more abundant in NGC 253 (Mauersberger & Henkel 1993; Jackson *et al.* 1995), indicating that NGC 253 has denser gas, and may be at an earlier stages of evolution than M82. Some of the "non-starbursts" in Figure 8 with rings of gas and star formation may be galaxies in an advanced post-starburst phase. However, since it is also possible to produce gas rings from dynamical effects (see §6.1), an understanding of both dynamical and evolutionary effects is essential for understanding circumnuclear regions.

Circumnuclear gas disks spawn circumnuclear starbursts which build compact circumnuclear stellar disks, and perhaps even stellar bulges. The central few hundred pc of many galaxies contain compact circumnuclear stellar disks, as revealed by spiral structure and rotational velocities which are large compared to their velocity dispersion (Kormendy 1993). Due to their compactness, some of these disks were previously thought to be the dense central portions of bulges. The starburst in NGC 3504 is in the process of making or adding to one of these compact stellar disks. Figure 10 shows that its central CO radial distribution is exponential with a scale length of 300 pc, similar to the radial distribution of K-band light (Kenney, Carlstrom, & Young 1993). Traditionally, the light from this compact component would be fit to an $R^{1/4}$ profile and called a bulge, but in NGC 3504 most of the light probably comes from a disk. The high gas surface density and relatively low CO velocity dispersion (2000 $M_\odot$ pc^{-2} and 20 km s^{-1} at $R = 200$ pc) imply the progenitor gas disk is thin. Recent simulations and theoretical work show that stars in an originally thin bar may be driven to large scale heights and form boxy or peanut-shaped bulges, either by instabilities (Raha *et al.* 1991; Sellwood 1993) or scattering at the vertical resonances of a bar (Combes *et al.* 1990; Pfenniger & Norman 1991). Similar mechanisms operating on circumnuclear stellar disks may contribute to spheroidal bulge building (Friedli & Benz 1995).

7. Vertical distribution of molecular gas

Molecular gas is the coldest and thinnest disk component in galaxies. In the Milky Way, throughout most of the plane ($R = 3 - 11$ kpc), the vertical distribution of molecular gas is similar to that of the OB stars, with a thickness (FWHM) of 90–150 pc (Scoville & Sanders 1987; Dame *et al.* 1987). The CO disk is thinner (FWHM = 65 pc) in the central 400 pc (Scoville & Sanders 1987) and thicker (FWHM = 200–350 pc) in the outer galaxy (Dame *et al.* 1987). It is generally assumed that the thickness of the gas layer is determined by a condition of hydrostatic equilibrium within a disk potential whose mass is mostly stars, although gas may contribute significantly in central regions. A thickening of the gas layer with increasing radius indicates that the disk mass surface density varies more strongly with radius than does the vertical gas velocity dispersion.

A nearby edge-on Sb galaxy which is remarkably similar to the Milky Way is NGC 891. Both galaxies have a nuclear molecular gas disk within $R \simeq 225$ pc and a molecular ring at $R = 3 - 7$ kpc (Garcia-Burillo *et al.* 1992; Scoville *et al.* 1993). The relative paucity of gas in between may be due to the action of stellar bars. Figure 11 shows an optical image of NGC 891, together with CO(1-0) maps at 2.3″ resolution from the OVRO interferometer (Scoville *et al.* 1993). Compared to the Milky Way, NGC 891 has about twice the CO scale height, and twice the molecular gas surface density. Since the two galaxies have very similar rotation curves and therefore similar mass distributions, the larger gas scale height implies a vertical velocity dispersion which is about twice as large in NGC 891. Interestingly, this means that the ratio between the gas surface density and the critical surface density for the onset of instability (Toomre 1964; Binney & Tremaine 1987) is about the same in the two galaxies (probably close to 1), which agrees with the concept of self-regulating star formation (e.g., Kennicutt 1989; Kennicutt, chapter 7).

While most of the molecular gas in NGC 891 is contained within a relatively thin layer (FWHM = 240 pc, Scoville *et al.* 1993), there may also be a thicker component. Single dish maps of CO(1-0) (21″ = 1 kpc resolution) and CO(2-1) (13″ = 600 pc resolution) from the IRAM 30m by Garcia-Burillo *et al.* (1992) show tantalizing evidence for CO(2-1) emission as far as 30″ = 1.4 kpc above the plane. Neither the single dish data nor the interferometer data show evidence for an extended component in the CO(1-0) distribution, implying that molecular gas in the halo, if truly present, is relatively hot and not too optically thick. Possible sidelobe problems for single dish data and missing flux problems for interferometer data mean that the molecular gas distribution in the halo of NGC 891 remains uncertain. Both atomic and ionized gas exist high in the halo of NGC 891,

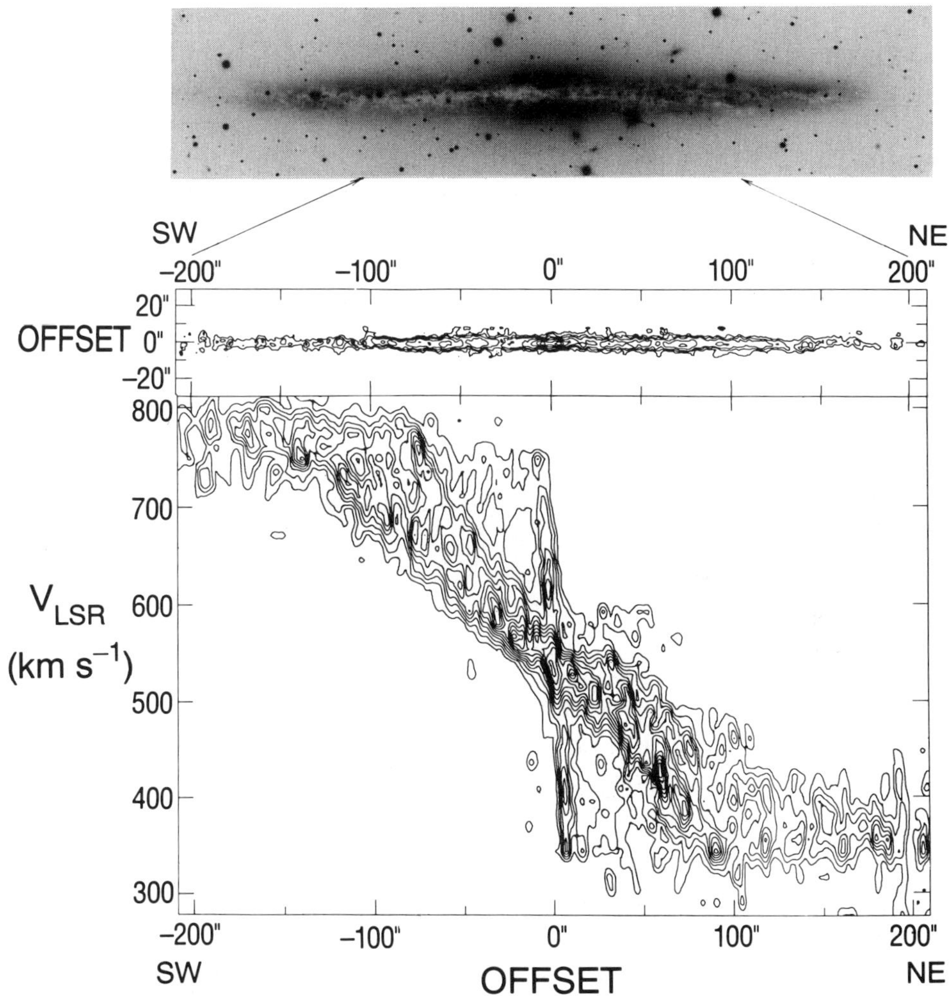

Figure 11. CO maps shown together with blue optical image of the edge-on Sb galaxy NGC 891. In the middle panel, CO emission is shown as a function of latitude and longitude, integrated over the full velocity range. In the lower panel, CO emission is shown as a function of longitude and velocity, integrated perpendicular to the galactic disk. The compact circumnuclear disk is a distinct feature in the longitude-velocity diagram. From Scoville *et al.* (1993).

as far as 5 kpc above the disk. This is presumably gas injected into the halo from the disk by the energy from star formation (see Rand, chapter 4). Knowledge of the molecular component of the halo is of interest for understanding the amount of gas in the halo, and the behavior of gas as it cycles between the disk and halo. Is there a significant molecular component to the gas being ejected from the disk? Does the gas cool sufficiently to become molecular as it settles toward the plane?

The nearly edge-on starburst galaxy M82 contains molecular gas at large distances from the plane. Both Nakai *et al.* (1987) and Yun (1992) find evidence for spur-like CO features extending from the central starburst region. The CO spurs lie roughly at the outer edge of the superwind of hot, outflowing gas, and may form the walls of the bipolar starburst outflow. From single dish data at $55'' = 880$ pc resolution, Yun (1992) finds a large extent (deconvolved FWHM = 820 pc) of CO(1-0) along the minor axis, which extends somewhat beyond the central starburst region. This large minor axis extent does not seem to be entirely due to gas lying in the plane (M82 is slightly inclined to the line of sight), and could be previously ejected gas re-settling into the disk.

In contrast to M82, most of the molecular gas fueling the young starburst in the nearly face-on spiral galaxy NGC 3504 seems to be confined to a thin disk. Assuming that the molecular gas is in hydrostatic equilibrium, the low CO velocity dispersions of 24–13 km s^{-1} at radii of 100–500 pc imply a disk thickness of only 20–40 pc (FWHM) (Kenney, Carlstrom, & Young 1993). The starburst in NGC 3504 seems to be in an earlier evolutionary phase than the one in M82, which may account for the difference. Perhaps only small amounts of gas have been ejected so far in NGC 3504, so that most of the gas still lies within a thin disk. Alternatively, the condition of hydrostatic equilibrium for molecular gas alone might not apply in the rapidly evolving, multi-phase interstellar media of starbursts.

8. Gas disks around active galactic nuclei

Gas which is not consumed by star formation or driven out by winds may reach the central few parsecs, where it may obscure or ultimately fuel an active nucleus. In this section we describe 2 active galaxies in which the relationship between the molecular gas and the nuclear activity is particularly clear.

8.1. NGC 1068

Seyfert 2 galaxies are believed to have much of their nuclear emission obscured by a dense torus of molecular gas (Antonucci & Miller 1985). Among the best observational evidence for dense molecular material being

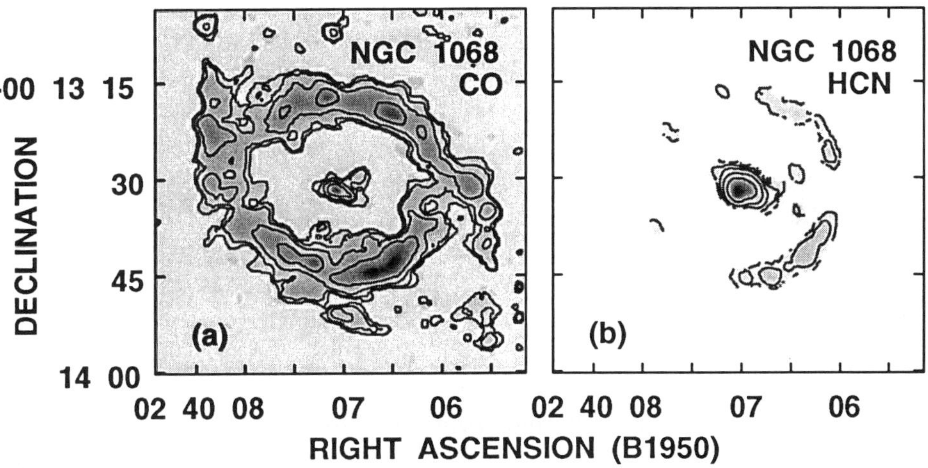

Figure 12. a.) CO (1-0) map of NGC 1068 at 2.9″ resolution obtained with Owens Valley interferometer (Planesas *et al.* 1991). Approximately half of the emission near the nucleus is continuum emission, so the true intensity of CO emission near the nucleus is even weaker than shown in this map. b.) HCN (1-0) map of NGC 1068 at 3.4×2.4″ resolution obtained with IRAM interferometer (Tacconi *et al.* 1994). While HCN emission is much weaker than CO emission in the molecular ring, the HCN/CO ratio is very high near the center, due to the influence of the active nucleus. CO emission appears weak near the nucleus because there is only a moderate mass of gas present. HCN emission appears strong because the gas that is there is all dense.

affected by, and perhaps obscuring an active nucleus is in NGC 1068, one of the nearest and best studied Seyfert 2 galaxies. Planesas *et al.* (1991) first detected CO emission from the central few hundred parsecs of NGC 1068, although the amount and distribution of molecular material was poorly constrained due to relatively strong millimeter continuum emission. Their map, reproduced in Fig. 12a, shows the peak CO brightness in the nucleus to be comparable to that in the prominent molecular ring at 15″ (1 kpc). The situation is totally different in maps of HCN (Jackson *et al.* 1993; Tacconi *et al.* 1994), which trace high density ($n \geq 10^5$ cm^{-3}) gas. While in CO (Fig 12a) the ring emission is stronger than the nuclear emission, the nucleus is dominant in the HCN map (Fig 12b). In the molecular ring the line intensity ratio is I(HCN)/I(CO)$\sim$0.01, whereas in the nucleus it is I(HCN)/I(CO)$\sim$0.5–2, which is by far the highest ratio ever observed in another galaxy. A line ratio of 1 means that virtually all the gas is dense ($n \geq 10^5$ cm^{-3}), and that there is almost no gas with moderate densities ($n \geq 10^3$ cm^{-3}), like that which provides most of the CO emission in nearby Milky Way clouds. In the central few hundred parsecs of most galaxies, the ratio I(HCN)/I(CO) ranges from 0.02--0.1 (Rieu *et al.* 1992), indicating that circumnuclear regions without active nuclei have a higher fraction of dense gas than the outer disks of galaxies, but still have most of their gas mass at moderate densities of 10^2 to 10^3 cm^{-3}. Jackson *et al.* (1993) suggest that moderate density cloud "envelopes" in the center of NGC 1068 have been stripped away or vaporized by the AGN, although the high ratio could also be due to a higher pressure in the intercloud medium. From line ratios, Tacconi *et al.* (1994) find that the molecular gas is warm (T_g>70 K), and from the large gas velocity dispersion (100 km s^{-1}), they argue that the molecular gas disk may be sufficiently thick to help obscure the nucleus Energy from the AGN could be responsible for heating the gas, thickening the gas disk, and stripping low density gas. The HCN kinematics imply rotation about a position angle of 33 or 45°(Jackson *et al.* 1993; Tacconi *et al.* 1994), consistent with the polar axis proposed for the obscuring torus (32°), although distinctly different from the rotation axis of the outer disk.

8.2. NGC 4258

Luminous H_2O masers originate from the central few pc of several nearby galaxies with some sort of nuclear activity, and can provide unique insight into the physical conditions very close to the nucleus (Greenhill *et al.* 1990). Among the best evidence for supermassive black holes in galactic centers comes from H_2O maser observations of the peculiar spiral galaxy NGC 4258, which has prominent and highly curved radio continuum jets, and a moderately active (LINER) nucleus. A spectrum of the H_2O line at 22 GHz, and a model of the source are shown in Fig. 13 (Greenhill *et al.* 1995). The spectrum shows numerous bright components near the systemic velocity of the galaxy, several components of moderate intensity redshifted by $\sim$800–1000 km s^{-1}, and a couple of weak components blueshifted by $\sim$900 km s^{-1} (Nakai *et al.* 1993; Miyoshi *et al.* 1995). VLBI observations of the masers show that the high-velocity features are offset from the systemic features in a nearly linear structure, and that velocity decreases with distance from the center in a Keplerian ($v \propto r^{-1/2}$) manner (Miyoshi *et al.* 1995). This provides compelling evidence that the maser emission originates from a nearly edge-on (83°) annulus of molecular gas with radii 0.1–0.2 pc, and rotating with a velocity of $\sim$1000 km s^{-1}. The individual spectral features near the systemic velocity drift in velocity over time, due to centripetal acceleration as the gas clumps orbit around the nucleus (Haschick, Baan, & Peng 1994). The measured drift rate, combined with a VLBI measurement of a linear gradient between maser position and velocity (Greenhill *et al.* 1995), yield a rotational speed of $\sim$900 km s^{-1}, and a radius of $\sim$0.1 pc, in excellent agreement with the observed redshifted and blueshifted maser features. The masers near the systemic velocity originate from the near part of the molecular disk, and amplify the radio continuum emission from the nucleus. The high-velocity features arise from the tangent points of the disk, where the path length is sufficiently long and the velocity gradient is sufficiently low that the maser emission is self-amplified. It is unclear why the blue lines are so much weaker than the red lines, although it could be due to a warp in the disk.

With a radius of 0.1 pc and a Keplerian velocity of 1000 km s^{-1}, the implied interior mass is 4×10^7 $M_\odot$. The density of 5×10^9 $M_\odot$ pc^{-3} is presently the highest dynamically determined density in a galaxy. A star cluster of this density would quickly collapse due to stellar collisions, making it very likely that the central mass is a black hole. The rotation axis of the molecular disk is parallel (in projection) to the inner kpc of the radio continuum jet, but differs by 119° from the outer disk rotation axis, indicating that the core is decoupled from the outer galaxy.

The masering disk is thin, appears warped, and probably precesses since the radio jets are curved. It is probably the inner edge of a molecular gas torus (Barvainis 1995). Such molecular tori are posited to explain the difference between Seyfert 1 and Seyfert 2 galaxies by obscuring the nuclear region of the more highly inclined Seyfert 2's (Antonucci & Miller 1985), but have never been clearly detected before. Although the masering portion of this disk seems too thin to provide much obscuration, perhaps the disk is sufficiently warped, or maybe the gas disk is thicker but only the thickest part mases, or maybe the disk is thicker at larger radii, where the conditions are unsuitable for masers. It is generally believed that much of the obscuration in Seyfert 2's occurs on 1 pc rather than 100 pc scales (Krolik and Begelman 1986), in part because there is significant emission from hot dust ($T \sim 1000$ K) in AGNs, meaning a rather substantial covering factor of dust very near in, at a pc or less (Barvainis 1987, 1990). On the other hand, the large HCN/CO ratio on $\sim$100 pc scales in NGC 1068 (Jackson *et al.* 1993; Tacconi *et al.* 1994), suggests that energy from the AGN affects molecular gas out to radii of 100 pc. This may be consistent with the idea that the mid- and far-IR radiation from AGN arises from dust which reprocesses the primary optical-UV radiation, and which lies within a (warped) disk extending to $\sim$100 pc (Sanders *et al.* 1989; Phinney 1989; Granato & Danese 1994).

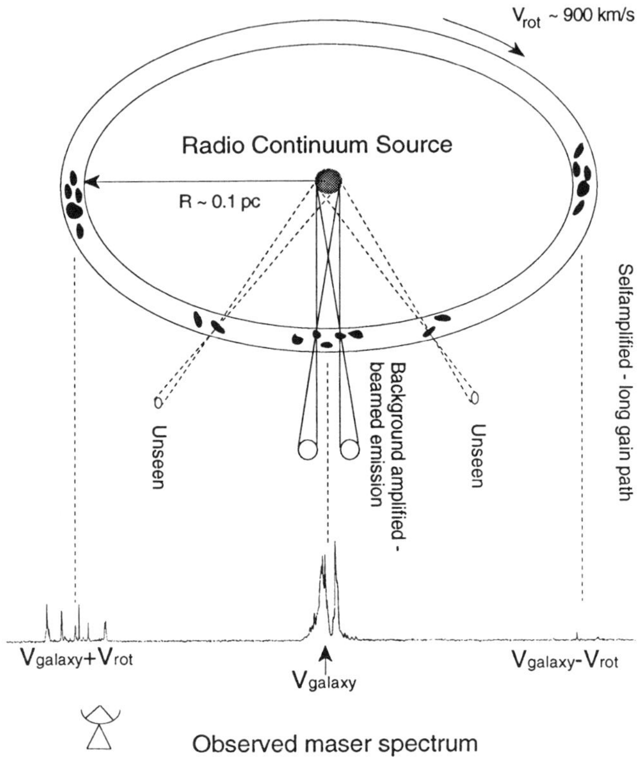

Figure 13. Observed H_2O maser spectrum in NGC 4258 and cartoon illustrating the rotating annulus model of the source. From Greenhill *et al.* (1995).

The high rotation velocities and small gas masses makes most molecular emission lines from these nuclear molecular disks broad and weak, and consequently difficult to detect. The strong and narrow lines of masers has made this discovery possible. Can we find other probes of the molecular gas tori around galactic nuclei in more highly inclined systems?

We thank Lincoln Greenhill, Nick Scoville, Linda Tacconi and Pere Planesas for providing figures or data to make figures, Mike Regan and Stu Vogel for providing material in advance of publication, and Rich Barvainis, Francoise Combes, Richard Larson, and Rich Rand for helpful comments.

References

Adler, D. S., Allen, R. J., and Lo, K. Y. 1991, *Ap.J.* , 382, 475.
Allen, R. J., Atherton, P. D., and Tilanus, R. P. J. 1986, *Nature* , 319, 296.
Allen, R. J., and Lequeux, J. 1993, *Ap.J.* , 410, L15.
Antonucci, R. R. J., and Miller, J. S. 1985, *Ap.J.* , 297, 621.
Arsenault, R., Boulesteix, J., Georgelin, Y., and Roy, J.-R. 1988, *Ast.Ap.* 200, 29.
Athanassoula, E. 1992, *Ast.Ap.* , 259, 345.
Barvainis, R. 1987, *Ap.J.* , 320, 537.
Barvainis, R. 1990, *Ap.J.* , 353, 419.
Barvainis, R. 1995, *Nature* , 373, 103.
Benedict, G. F., *et al.* 1993, *A.J.* , 105, 1369.
Benedict, G. F., Smith, B., and Kenney, J. D. P. 1996, *A.J.* , 111, 1861.
Benedict, G. F., Smith, B., and Kenney, J. D. P. 1996, *A.J.* , 112, 1318.
Binney, J., and Tremaine, S. 1987, *Galactic Dynamics*, (Princeton: Princeton University Press).
Blitz, L., Binney, J., Lo, K.-Y., Bally, J., and Ho, P. T. P. 1993, *Nature* , 361, 417.
Bloemen, J. B. G. M. 1989, *Ann.Rev.Ast.Ap.* , 27, 469.
Boer, B. 1994, in *Mass-Transfer Induced Activity in Galaxies*, ed. I. Shlosman, (Cambridge: Cambridge University Press), p. 308.
Bosma, A. 1981, *A.J.* , 86, 1971.
Broeils, A. 1992, PhD thesis, Groningen.
Carlstrom, J. E. 1988, in *Galactic and Extragalactic Star Formation* eds. R. E. Pudritz and M. Fich, (Dordrecht: Kluwer), p. 571.
Combes, F. 1988, in *Galactic and Extragalactic Star Formation* eds. R. E. Pudritz and M. Fich, (Dordrecht: Kluwer), p. 475.
Combes, F. 1991, *Ann.Rev.Ast.Ap.* , 29, 195.
Combes, F., and Elmegreen, B. G. 1993, *Ast.Ap.* , 271, 391.
Combes, F., and Gerin, M. 1985, *Ast.Ap.* , 150, 327.
Combes, F., Gerin, M., Nakai, N., Kawabe, R., and Shaw, M. A. 1992, *Ast.Ap.* , 259, L27.
Combes, F., Debbasch, F., Friedli, D., and Pfenniger, D. 1990, *Ast.Ap.* , 233, 82.
Dame, T. M., *et al.* 1987, *Ap.J.* , 322 706.
Dame, T. M., Koper, E., Israel, F. P., and Thaddeus, P. 1993, *Ap.J.* , 418, 730.
de Geus, E. J., and Phillips, J. A. 1996, in IAU Symp. 169, *Unsolved Problems of the Milky Way*, ed. L. Blitz, (Dordrecht: Kluwer), p. 575.
Deutsch, E. W., and Allen, R. J. 1993, *A.J.* , 106, 1812.
Dickman, R. L., Snell, R. L., and Schloerb, F. P. 1986, *Ap.J.* , 309, 326.
Digel, S., de Geus, E.J. , and Thaddeus, P. 1994, *Ap.J.* , 422, 92.
Elmegreen, B. G. 1979, *Ap.J.* , 231, 372.
Elmegreen, B. G. 1988, *Ap.J.* , 326, 616.
Elmegreen, B. G. 1989, *Ap.J.* , 338, 178.
Elmegreen, B. G. 1993, *Ap.J.* , 411, 170.
Elmegreen, B. G. 1994, *Ap.J.* , 425, L73.
Elmegreen, B. G., and Elmegreen, D. M. 1985, *Ap.J.* , 288, 438.
Falgarone, E., Puget, J.-L., and Perault, M. 1992, *Ast.Ap.* , 257, 715.
Federman, S. R., Glassgold, A. E., and Kwan, J. 1979, *Ap.J.* , 227, 466.
Freeman, K. C. 1992, in *Physics or Nearby Galaxies: Nature or Nurture?*, eds. T. X. Thuan, C. Balkowski, and J. T. T. Van, (Gif-sur-Yvette: Editions Frontieres), p. 201.
Friedli, D., and Benz, W. 1995, *Ast.Ap.* , 301, 649.
Garcia-Burillo, S., Combes, F., and Gerin, M. 1993b, *Ast.Ap.* , 274, 148.
Garcia-Burillo, S., Guelin, M., Cernicharo, J., and Dahlem, M. 1992, *Ast.Ap.* , 266, 21.
Garcia-Burillo, S., Guelin, M., and Cernicharo, J. 1993a, *Ast.Ap.* , 274, 123.
Granato, G. L., and Danese, L. 1994, *M.N.R.A.S.* , 268, 235.
Greenhill, L. J., Jiang, D. R., Moran, J. M., and Reid, M. J. 1995, *Ap.J.* , 440, 619.

Greenhill, L. J., Moran, J. M., Reid, M. J., Gwinn, C. R., Menten, K. M., Eckart, A., and Hirabayashi, H. 1990, *Ap.J.* , 364, 513.

Handa, T., Nakai, N., Sofue, Y., Hayashi, M., and Fujimoto, M. 1990, *P.A.S.J.* , 42, 1.

Handa, T., Sofue, Y., and Nakai, N. 1991, in *Dynamics of Galaxies and Their Molecular Cloud Distributions*, IAU Symp. 146, ed. F. Combes and F. Casoli (Dordrecht: Kluwer), p. 156.

Haschick, A. D., Baan, W. A., and Peng, E. W. 1994, *Ap.J.* , 437, L35.

Heckman, T. M., Armus, L., and Miley, G. K. 1990, *Ap.J.Suppl.* , 74, 833.

Helfer, T. T., and Blitz, L. 1993, *Ap.J.* , 419, 86.

Heller, C. H., and Shlosman, I. 1994, *Ap.J.* , 424, 84.

Henkel, C., Baan, W. A., and Mauersberger, R. 1991, *Ast.Ap.Rev.* , 3, 47.

Hernquist, L., and Mihos, J. C. 1995, *Ap.J.* , 000, 000***

Hurt, R. L., and Turner, J. L. 1991, *Ap.J.* , 377, 434.

Ishizuki, S. 1994, in *Astronomy with Millimeter and Submillimeter Wave Interferometry*, eds. M. Ishiguro and W. J. Welch, (San Francisco: Astronomical Society of the Pacific), p. 293.

Ishizuki, S., Kawabe, R., Ishiguro, M., Okumura, S. K., Morita, K.-I., Chikada, Y., and Kasuga, T. 1990, *Nature* , 344, 224.

Israel, F. P., *et al.* 1993, *Ast.Ap.* , 276 25.

Jackson, J. M., Paglione, T. A. D., Carlstrom, J. E., and Rieu, N. Q. 1995, *Ap.J.* , 438, 695.

Jackson, J. M., Paglione, T. A. D., Ishizuki, S., and Rieu, N. Q. 1993, *Ap.J.* , 418, L13.

Jogee, S., and Kenney, J. D. P. 1995, *Bull. Am. Ast. Soc.* 187, 4906.

Jogee, S., and Kenney, J. D. P. 1997, in preparation

Keel, W. C. 1993, *A.J.* , 106, 1771.

Kenney, J. D. P., Carlstrom, J. E., and Young, J. S. 1993, *Ap.J.* , 418, 687.

Kenney, J. D. P., and Lord, S. D. 1991, *Ap.J.* , 381, 118.

Kenney, J. D. P., Wilson, C. D., Scoville, N. Z., Devereux, N. A., and Young, J. S. 1992, *Ap.J.* , 395, L79.

Kenney, J. D. P., Wilson, C. D., Pisano, D. J., and Scoville, N. Z., 1997, *Ap.J.* , submitted

Kenney, J. D. P., and Young, J. S. 1988, *Ap.J.* , 326, 588.

Kenney, J. D. P., and Young, J. S. 1989, *Ap.J.* , 344, 171.

Kennicutt, R. C., Jr. 1989, *Ap.J.* , 344, 685.

Knapen, J. H., Beckman, J. E., Shlosman, I., Peletier, R. F., Heller, C. H., and de Jong, R. S. 1995, *Ap.J.* , 443, L73.

Koribalski, B., Dickey, J. M., and Mebold, U. 1993, *Ap.J.* , 402, L41.

Kormendy, J. 1993, in *Galactic Bulges*, ed. H. Dejonge and H. J. Habing, (Dordrecht: Kluwer), p. 209.

Krolik, J. H., and Begelman, M. C. 1986, *Ap.J.* , 308, L55.

Larson, R. B. 1981, *M.N.R.A.S.* , 194, 809.

Lequeux, J., Allen, R. J., and Guilloteau, S. 1993, *Ast.Ap.* , 280, L23.

Lequeux, J., Le Bourlot, J., Pineau des Forets, Roueff, E., Boulanger, F., and Rubio, M. 1994, *Ast.Ap.* , 292, 371.

Lesch, H., Biermann, P. L., Crusius, A., Reuter, H. P., Dahlem, M., Barteldress, A., and Wielebinski, R. 1990, *M.N.R.A.S.* , 242, 194.

Lindqvist, M., Winnberg, A., Habing, H. J., and Matthews, H. E. 1991, *Ast.Ap.* Supp., 92, 43.

Lord, S. D., and Kenney, J. D. P. 1991, *Ap.J.* , 381, 130.

Lord, S. D., and Young, J. S. 1990, *Ap.J.* , 356, 135.

Magnani, L., Blitz, L., and Mundy, L. 1995, *Ap.J.* , 295, 402.

Magnani, L., and Onello, J. S. 1995, *Ap.J.* , 443, 169.

Maloney, P., and Black, J. H. 1988, *Ap.J.* , 325, 389.

Martin, P., and Roy, J.-R. 1995, *Ap.J.* , 445, 161.

Mauersberger, R., and Henkel, C. 1993, Reviews in Modern Astronomy, 6, 69.

Mead, K. N., and Kutner, M. L. 1988, *Ap.J.* , 330, 399.

Mihos, J. C., and Hernquist, L. 1994, *Ap.J.* , 425, L13.

Miyoshi, M., Moran, J., Herrnstein, J., Greenhill, L., Nakai, N., Diamond, P., and Inoue, M. 1995, *Nature* , 373, 127.

Nakai, N., Hayashi, M., Handa, T., Sofue, Y., and Hasegawa, T. 1987, *P.A.S.J.* , 39, 685.

Nakai, N., Inoue, M., and Miyoshi, M. 1993, *Nature* , 361, 45.

Oka, T., Hasegawa, T., Handa, T., Hayashi, M., and Sakamoto, S. 1995, *Ap.J.* , 000, 000***

Pauls, T., and Mezger, P. G. 1975, *Ast.Ap.* , 44, 259.

Phinney, E. S. 1989, in *Theory of Accretion Disks*, ed. Meyer, J. S., (Kluwer: Dordrecht), p. 457.

Pfenniger, D., and Combes, F. 1994, *Ast.Ap.* , 285, 94.

Pfenniger, D., Combes, F., and Martinet, L. 1994, *Ast.Ap.* , 285, 79.

Pfenniger, D., and Norman, C. 1990, *Ap.J.* , 363, 391.

Planesas, P., Scoville, N. Z., and Myers, S. T. 1991, *Ap.J.* , 369, 364.

Polk, K. S., Knapp, G. R., Stark, A. A., and Wilson, R. W. 1988, *Ap.J.* , 332, 432.

Quillen, A. C., Frogel, J. A., Kenney, J. D. P., Pogge, R. W., and DePoy, D. L. 1995, *Ap.J.* , 441, 549.

Raha, N., Sellwood, J. A., James, R. A., and Kahn, F. D. 1991, *Nature* , 352, 411.

Rand, R. J. 1993a, *Ap.J.* , 404, 593.

Rand, R. J. 1993b, *Ap.J.* , 410, 68.

Rand, R. J. 1995, *A.J.* , 109, 2444.

Rand, R. J., and Kulkarni, S. R. 1990, *Ap.J.* , 349, L43.

Regan, M. W., and Vogel, S. N. 1995, *Ap.J.* , 452, 21.

Regan, M. W., Vogel, S. N., and Teuben, P. 1995, *Ap.J.* , 449, 596.

Rieu, N. Q., Jackson, J. M., Henkel, C., Bach, T., and Mauersberger, R. 1992, *Ap.J.* , 399, 521.

Rix, H.-W., and Rieke, M. J. 1993, *Ap.J.* , 418, 123.

Roberts, W. W., Jr., Huntley, J. M., and van Albada, G. D. 1979, *Ap.J.* , 233, 67.

Roberts, W. W., Jr., and Stewart, G. R. 1987, *Ap.J.* , 314, 10.

Rubio, M., Lequeux, J., and Boulanger, F. 1993, *Ast.Ap.* , 271 9.

Sakamoto, K., Okumura, S., Minezaki, T., Kobayashi, Y., and Wada, K. 1995, *A.J.* , 110, 2075.

Sanders, D. B., Phinney, E. S., Neugebauer, G., Soifer, B. T., and Matthews, K. 1989, *Ap.J.* , 347, 29.

Sanders, R. H., Wrixon, G. T., and Mebold, U. 1977, *Ast.Ap.* , 61, 329.

Scoville, N. Z., Hibbard, J. E., Yun, M. S., and van Gorkom, J. H. 1994, in *Mass-Transfer Induced Activity in Galaxies*, ed. I. Shlosman, (Cambridge: Cambridge University Press), p. 191.

Scoville, N. Z., Sargent, A. I., Sanders, D. B., and Soifer, B. T., 1991, *Ap.J.* , 366, L5.

Scoville, N. Z., Thakkar, D., Carlstrom, J. E., and Sargent, A. I., 1993, *Ap.J.* , 404, L59.

Scoville, N. Z., and Sanders, D. B. 1987, in *Interstellar Processes*, ed. H. Thronson and D. Hollenbach (Dordrecht: Reidel), p. 21.

Sellwood, J. A. 1994, in *Numerical Simulations in Astrophysics*, ed. J. Franco, S. Lisano, L. Aguilar and E. Daltabuit (Cambridge: Cambridge University Press). p. 90.

Shier, L. M., Rieke, M. J., and Rieke, G. H. 1994, *Ap.J.* , 433, L9.

Shlosman, I., Frank, J., and Begelman, M. C. 1989, *Nature* , 338, 45.

Smith, P. A., Brand, P. W. J. L., Puxley, P. J., Mountain, C. M., and Nakai, N. 1990, *M.N.R.A.S.* , 243 97.

Sodroski, T. J. 1991, *Ap.J.* , 366, 95.

Solomon, P. M., Rivolo, A. R., Barrett, J., and Yahil, A. 1987, *Ap.J.* , 319 730.

Stark, A. A., Gerhard, O. E., Binney, J., and Bally, J. 1991, *M.N.R.A.S.* , 248, 14P.

Strong, A. W., Bloemen, J. B. G. M., Dame, T. M., Grenier, I. A., Hermsen, W., Lebrun, F., Nyman, L.-A., Pollock, A. M. T., and Thaddeus, P. 1988, *Ast.Ap.* , 207, 1.

Tacconi, L. J., Genzel, R., Blietz, M., Cameron, M., Harris, A. I., and Madden, S. 1994, *Ap.J.* , 426, L77.

Toomre, A. 1964, *Ap.J.* , 139, 1217.
Tubbs, A. D. 1982, *Ap.J.* , 255, 458.
Turner, J. L. 1994, in *Mass-Transfer Induced Activity in Galaxies*, ed. I. Shlosman, (Cambridge: Cambridge University Press), p. 90.
Turner, J. L., and Hurt, R. L. 1992, *Ap.J.* , 384, 72.
Vogel, S. N., Kulkarni, S. R., and Scoville, N. Z. 1988, *Nature* , 334, 402.
Wada, K., and Habe, A. 1992, *M.N.R.A.S.* , 259, 82.
Wada, K. 1994, *P.A.S.J.* , 46, 165.
Wiklind, T., Rydbeck, G., Hjalmarson, A., and Bergman, P. 1990, *Ast.Ap.* , 232, L11.
Wild, W., Harris, A. I., Eckart, A., Genzel, R., Graf, U. U., Jackson, J. M., Russell, A. P. G., and Stutzki, J. 1992, *Ast.Ap.* , 265 447.
Wilson, C. D. 1994, in *The Local Group: Comparative and Global Properties*, eds. A. Layden, C. Smith, and J. Storm, CTIO/ESO workshop proceedings, p. 15.
Wilson, C. D. and Rudolph, A. L. 1993, *Ap.J.* , 406, 477.
Wilson, C. D. and Scoville, N. 1990, *Ap.J.* , 363, 435.
Wilson, C. D. and Scoville, N. 1992, *Ap.J.* , 385, 512.
Wozniak, H., Friedli, D., Martinet, L., Martin, P., and Bratschi, P. 1995 *Ast.Ap.Suppl.* , 111, 115.
Young, J. S., Xie, S., Kenney, J. D., and Rice, W. L. 1989, *Ap.J.Suppl.* , 70, 699.
Young, J. S., and Knezek, P. M. 1989, *Ap.J.* , 347, L55.
Young, J. S., and Scoville, N. Z. 1982, *Ap.J.* , 258, 467.
Young, J. S., and Scoville, N. Z. 1991, *Ann.Rev.Ast.Ap.* , 29, 581.
Yun, M. S. 1992, Ph.D. thesis, Harvard University

COOL H I DISKS IN GALAXIES

M.P. HAYNES AND A.H. BROEILS
Center for Radiophysics and Space Research and
National Astronomy and Ionosphere Center,
Cornell University,
Space Sciences Building, Ithaca,
NY 14853
U.S.A.

1. Introduction

Because of the sensitivity and resolution, both spatial and spectral, at centimeter wavelengths and the large amounts of interstellar gas in the form of neutral atomic hydrogen (H I) in typical galaxies, the 21-cm H I line has served as a valuable tool in probing the disk component of galaxies, both structurally and dynamically. Even without spatial information, the global 21-cm profile provides three important parameters: the redshift, the total H I line flux, and line width, the latter a measure of the Doppler broadening. The use of the H I line for redshift surveys has been discussed by Giovanelli & Haynes (1990); H I redshifts remain among the most accurate determinations available to redshift surveys and have been measured for more than 12000 galaxies. The total H I line flux is used as a measure of the potential for future star formation and as an indicator of past gas removal. Spatial mapping of the H I distribution allows the intercomparison of the gas and stellar distributions. The global H I line width in rotationally dominated galaxies has been used via the Tully–Fisher relation (e.g. Rowan–Robinson 1985) as a predictor of galaxy luminosity, under the assumptions that rotation speed predicts mass and mass predicts luminosity. Mapping of the velocity field provides an indication of the distribution of mass, both luminous and dark. Thus the study of the H I in galaxies addresses a large number of issues associated with galaxy structure and formation and also provides clues to their distribution and evolution.

In the Milky Way, the detail of the H I distribution has not only revealed

J. M. van der Hulst (ed.), The Interstellar Medium in Galaxies, 75–103.

the spiral structure, but also the plethora of exotic phenomena: the intermediate and high velocity clouds, the so–called Lockman layer, the wormholes and chimneys, etc. The various latter phenomena, representative of the more energetic phenomena seen also in other galaxies, lie in the purview of chapter 4 by Richard Rand to which the reader is referred for their discussion. Here we focus on the details of the cool H I layer and in particular, its relationship to current star formation and its use as a tracer of galactic structure and as an indirect probe of a galaxy's evolutionary history.

In the interest of brevity, this chapter must necessarily focus on illustrative examples rather than being all encompassing. We attempt to refer the interested reader to other important sources, especially reviews, and apologize for not being able to recognize the important contributions of many.

Two recent and thorough reviews of the distribution of H I in the Milky Way have been presented by Burton (1988) and by Dickey & Lockman (1990). Giovanelli & Haynes (1988) provide a review of the methodology behind 21-cm line observations. A good summary of similar discussion of the H I distribution in galaxies was given previously by Sancisi (1980).

2. Neutral Hydrogen Content and Global Properties

The H I line flux when measured over the entire solid angle occupied by the galaxy gives a direct measure of the total mass of H I $\mathcal{M}_{\rm HI}$. Currently H I line fluxes, or upper limits, are available for some 15000 galaxies skywide. The global H I content of galaxies has been the subject of recent reviews by Haynes *et al.* (1984) and Giovanelli & Haynes (1988). Roberts & Haynes (1994; hereafter RH94) present a review of the variation of a variety of physical properties of galaxies, include the H I content, along the Hubble sequence. Our discussion of the global properties is based on RH94 to which the reader is referred for more detail.

The detected H I line flux $\int SdV$ must in practice be corrected for possible beam effects when single dish radio telescopes are used to measure it. As discussed by Roberts (1975), the total H I mass is derived from the correct H I line flux by the simple expression $\mathcal{M}_{\rm HI} = 2.36 \times 10^5\ D^2 \int SdV$, where D is the distance in Mpc and $\int SdV$ is given in Jy kms^{-1}. Use of this conversion from flux to mass assumes that the H I clouds are optically thin; in denser regions of the interstellar medium, this assumption does not hold (see Section 3.2). Where the H I line is not detected but a measure of the rms noise per unit velocity is available, an upper limit to the H I flux can be estimated from the noise limit, assuming some distribution over velocity. RH94 assume a box–like H I line profile of amplitude 1.5 times the rms noise and width as predicted for each morphological class by the luminosity–velocity width relation derived for the detected objects; the flux from a profile of roughly these characteristics would have been measurable. Non–detections comprise a significant fraction of the samples of early type galaxies E–Sab and are insignificant for later types. RH94 performed an analysis of two large samples extracted from the Third Reference Catalog of Galaxies (de Vaucouleurs *et al.* 1991) one of which is limited to the volume occupied by the Local Supercluster. For the study of H I content, this nearby sample is most useful because of its high degree of completeness, even among the early–type galaxies.

Not surprising, $\mathcal{M}_{\rm HI}$ generally increases as galaxies become bigger and more luminous, spanning a range of over 4 orders of magnitude. Because of the desire to disentangle parameters that vary only according to the size or brightness of a galaxy (parameters of scale) from those that depend on morphological or other properties, several other quantities are generally used in analyzing the total H I content of galaxies. The most useful indicators are the hydrogen mass to luminosity ratio $\mathcal{M}_{\rm HI}/\mathcal{L}_{\rm B}$, the H I mass fraction $\mathcal{M}_{\rm HI}/\mathcal{M}_{\rm T}$ and the global H I surface density $\sigma_{\rm HI}$. $\mathcal{M}_{\rm HI}/\mathcal{L}_{\rm B}$ is most useful as a comparative indicator within a Hubble class and when luminosity bias is not an issue. For most galaxies, $\mathcal{M}_{\rm HI}/\mathcal{L}_{\rm B}$ is of order a few tenths (in solar units). For statistical purposes, the total mass used to calculate $\mathcal{M}_{\rm HI}/\mathcal{M}_{\rm T}$ is limited to the luminous portion of the galaxy $\mathcal{M}_{\rm T}(< R_{25})$. It is well known that the fractional H I mass $\mathcal{M}_{\rm HI}/\mathcal{M}_{\rm T}$ increases along the Hubble sequence: E and S0 galaxies contain proportionately lower H I masses whereas for the latest Sm's and Im's, the H I contributes several tenths of the total galaxy mass within the optical radius. As is discussed in Section 4.1, the true total mass (luminous + dark) may be considerably larger than $\mathcal{M}_{\rm T}(< R_{25})$ so that, especially for the latest types, the overall fractional H I mass probably does not grow much. The details of the H I distributions along the morphological sequence and the relationship of the H I component to the underlying old stars, sites of on–going star formation and the dark matter distribution are discussed in later sections.

The globally averaged H I surface density $\sigma_{\rm HI}$ is a useful measure of the H I content especially when interpreted in terms of the potential for future star formation. As complements to Figures 3 and 4 and Table 1 of RH94, Figure 1 displays the detailed histograms for subgroups of morphological class of $\sigma_{\rm HI}$, in logarithmic units of $\rm M_{\odot}\,pc^{-2}$, for the Local Supercluster sample of RH94. It should be noted that this global property is a hybrid quantity because the optical diameter (specifically D_{25}) has been used as the area indicator. Cross–hatched areas of the histograms indicate galaxies not detected in the H I line so that the estimated H I mass is an upper limit. It is clear from Figure 1 that the early types show a much larger range of $\sigma_{\rm HI}$ than do the later spirals. While some E's and S0's have H I contents similar to those of Sb–Sc spirals, others contain several orders of magnitude less

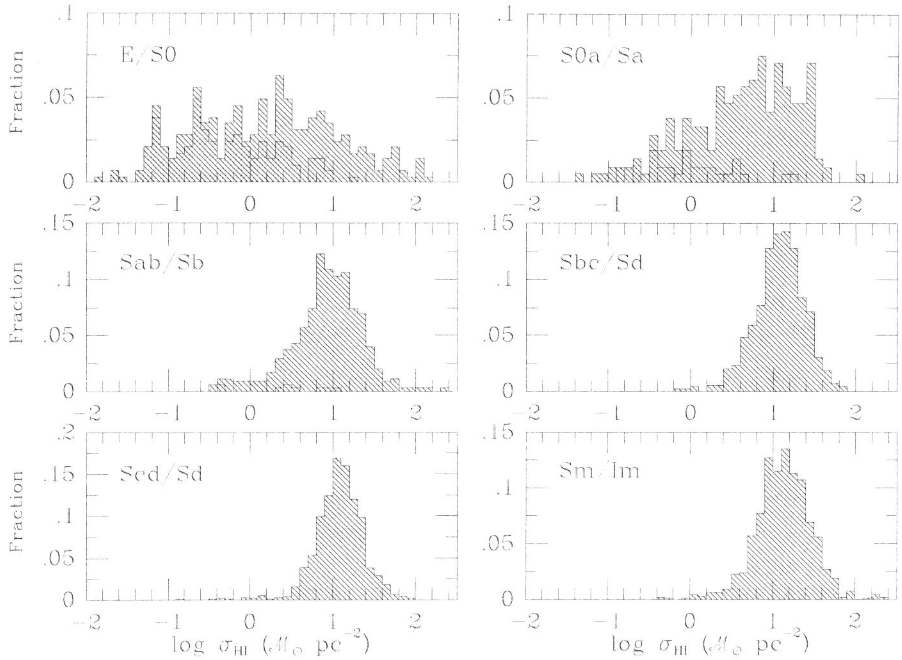

Figure 1. Normalized histograms of the HI surface density σ_{HI}, in logarithmic units of $M_\odot\,pc^{-2}$, for galaxies of different morphologies in the Local Supercluster sample. Cross hatched regions indicate upper limits derived from non–detections. After RH94.

H I . The great dispersion in H I contents among the early types is discussed in Section 5.2. Among the later–type spirals, σ_{HI} is useful as an indirect probe of the effect of local environment on star formation potential. The effect of environment is discussed further in Section 6.2.

In the following sections, we summarize the details of the H I distribution and the velocity field extracted from H I synthesis observations to explore the structure of H I disks in galaxies, their origin and evolution.

3. Distribution of H I in Disks

Detailed studies of the H I distribution in galaxies at moderate spatial resolutions are available for several dozens of galaxies. Isolated spiral galaxies have in general regular, extended H I distributions, although many disks show some lopsidedness. The spiral arm structure can often been recognized in the H I and in some cases can be followed beyond the optical radius. Here we summarize the main properties of the H I distributions gleaned from the mapping of nearby galaxies.

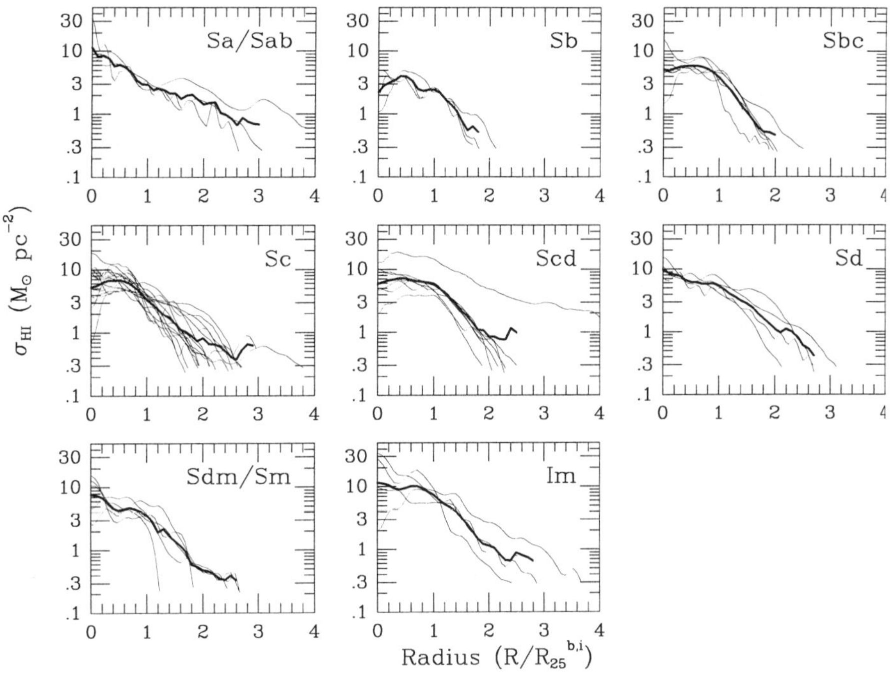

Figure 2. Radial surface density profiles of neutral hydrogen for eight spiral sub-classes. The radius is normalized by the face-on optical radius at the 25th mag arcsec^{-2} isophote. The thick solid line is the average profile for each type. From Broeils & van Woerden (1994).

3.1. RADIAL DISTRIBUTION

The H I extent of a galaxy is typically larger than the optical radius. Indeed, in the Milky Way, the radial distribution of H I is quite different from that seen in other Pop I tracers. The radial distribution of H I is depressed inwards of 4 kpc from the center, is relatively flat over the disk from 4 to 14 kpc, and then declines toward the outside. Most spirals have similar radial distributions — flat over the optical region and a sharp drop beyond the optical edge; in types later than Sc the density profiles generally continue to rise all the way into the center, reaching higher central values than in the other types. Figure 2 shows a selection of radial surface density profiles $\sigma_{HI}(R)$, ordered by morphological class, from Broeils & van Woerden (1994).

In the Milky Way, the H I extends to at least 25 kpc, farther out than any other disk tracer. H I disks in external galaxies range in sizes from a few kiloparsecs up to about a 100 kpc in isolated galaxies to a few hundred kpc in interacting systems. The spread in the ratio of H I diameter (at a level of $1\,M_\odot\,pc^{-2}$) to inclination–corrected optical diameter ($R_{25}^{b,i}$) is however remarkable small: For a sample of 50 isolated galaxies Broeils &

van Woerden find an average of 1.8 ± 0.4. Typically H I disks appear to cut off sharply at a density level of about 10^{19} atoms cm^{-2} $\sim 0.1\,M_{\odot}\,pc^{-2}$. A very deep map of the Sc galaxy NGC 3198 obtained by van Gorkom *et al.* (1993, described by van Gorkom 1992) after about a 100 hours of integration with the VLA indicates that its H I disk has a very sharp edge: the H I column density drops from 2×10^{19} atoms cm^{-2} to 4×10^{18} atoms cm^{-2} (= 3σ detection limit) within one synthesized beam of 2.7 kpc. Corbelli *et al.* (1989) found a similar abrupt truncation in M33 with sensitive Arecibo observations. The most likely explanation for this phenomena is that the neutral portion of the gas below a critical column density gets rapidly ionized by the intergalactic UV radiation field (Maloney 1993; Corbelli & Salpeter 1993). This suggests that galaxies have even more extended gas (H I + H II) envelopes than detected by current 21-cm line observations. Attempts have been made to detect such envelopes using H I absorption measurements (e.g. Carilli & van Gorkom 1992) and Lyα absorption systems (Bahcall *et al.* 1991; Morris *et al.* 1991). More details can be found in van Gorkom (1992) for the observational aspects and in Salpeter (1995 and references therein) for theoretical explanations.

3.2. WARM AND COOL GAS

Although the title of this chapter is "cool H I disks in galaxies", it is important to realize that atomic hydrogen in galaxies can be found in two very distinct phases, generally referred to as 'warm' and 'cool' gas (see review by Kulkarni & Heiles 1988). In the Milky Way, about 25% of the H I is in the cool phase and 75% in the warm, although this ratio may depend on location. The relative abundances of the two phases can be determined by comparing emission and absorption in the 21-cm line (cf. Dickey & Lockman 1990). Absorption in the 21-cm line occurs only in the cool gas, because the absorption coefficient κ is inversely proportional to the temperature. H I absorption experiments in M31 and M33 (Braun & Walterbos 1992; Dickey & Brinks 1993) have revealed that in those external galaxies also, a considerable fraction (40% and 15% respectively) of the atomic hydrogen is in the form of cool ($\sim$ 100 K) gas. Furthermore, for M31 a radial temperature gradient is found in the cool component: from $\sim$ 75 K in the central regions to about 175 K further out.

The H I structure in other galaxies, if observed at high enough resolution, is very complex and filamentary. A detailed account of the variety of structures and a discussion of the interaction between star formation (stellar winds and supernova explosions) can be found in van der Hulst (1996) and references therein.

Recent very high spatial ($\sim$ 100 pc) and velocity (6 km s^{-1}) resolution observations of several nearby galaxies by Braun (1995) have shown that also the H I *emission* from disks can be resolved into two distinct components: a high–brightness filamentary network (so–called HiBFiN) with transverse sizes of about 100 pc and velocity widths of less than 6 km s^{-1} , and a diffuse disk of low surface brightness. The suggestion is that the HiBFiN is in fact the cool condensed phase of the ISM with large local opacities, and that it corresponds spatially with starforming regions in spiral arms. On average, each component contributes about half of the total line flux; Braun estimates however that the total mass in the filamentary structures may exceed the diffuse component by a significant factor. For the galaxies in his sample, he finds that the H I masses calculated using the common assumption that H I emission is an optically thin tracer of the H I mass (Section 2) are low by factors between 1.1 and 2.2, with an average of 1.3 (Braun, private communication; Braun, 1997).

3.3. WARPS, FLARES AND THE Z–DISTRIBUTION OF H I

While the gas inside the solar circle defines well a flat disk of small scale height, the outer H I displays both a flare and a warp, as seen clearly in the maps of Burton & te Linkel Hekkert (1986). Beginning just beyond the solar circle, the warp is not symmetric, rising more in the northern hemisphere than in the south. The thickness of the H I layer in the inner Galaxy is relatively constant, of half thickness approximately 220 pc (Lockman 1983). Although most of the H I is confined to the inner half thickness, a thicker disk, referred to as the Lockman layer, also appears present, of half thickness roughly twice as large. The gas in the Lockman layer has a higher average temperature than that in the thin disk, is more smoothly distributed and thus has a larger filling factor, and a higher characteristic velocity dispersion (about 11 km s^{-1} rather than 6–8 km s^{-1} in the thin disk).

Flaring gas layers in external galaxies are not well studied because they require high resolution, high sensitivity H I observations of edge–on galaxies. Besides, detailed modeling of the gas distribution is often necessary to determine the position of gas clouds in the galactic disk along the long line-of-sight through the edge–on disk. flares are found in, among others, NGC 891 (Sancisi & Allen 1979; van der Kruit 1981, Swaters *et al.* 1995, 1997), NGC 100, UGC 9242 (Bosma 1991) and NGC 4244 (Olling & van Gorkom 1995). The increasing thickness of an H I layer with radius can be used to trace the radial behavior of the vertical force field in order to attempt to constrain the shape of dark halos. This is discussed in more detail in Section 4.2.

In NGC 891 the vertical H I extent appears associated with the star formation activity in the disk (Swaters *et al.* 1995, 1997; see also Rand, chapter 4). Swaters *et al.* find that the H I above the plane is rotating slower than the H I in the disk. This may reflect the precise mass distribution.

Warps in the H I layers of galactic disks are very common: Bosma (1991) estimates that nearly 50% of all H I disks are warped. Warps can be detected in two ways. In edge–on galaxies bending of the gas layer from the principal plane can be seen directly. The most extreme example is the prodigious warp in NGC 4013 (Bottema *et al.* 1987; Bottema 1995, 1996). Most warps are however inferred from the kinematics and can be modeled well by the tilted ring approach (Bosma 1981a,b; Begeman 1987; Broeils 1992a, cf. Section 4.1). Although warps in galaxies have been well studied, no consensus has been reached on a complete theory that explains the high percentage of warps, their origin, and their perseverance against smearing due to precession (reviewed by Binney 1992). Briggs (1990) has analyzed sixteen galaxies with significant warps and stated a couple of "rules" which hold for twelve of them. He finds that a warp usually starts at R_{25} and that, if viewed from the inner plane, it has a straight line of nodes between R_{25} and the Holmberg radius. Further out, where the warping amplitude is largest, the line of nodes winds in the leading sense. Within certain theoretical frameworks the warping of H I disks can also be used to constrain the halo flattening (cf. Sparke & Casertano 1988).

3.4. COMPLEX VELOCITY FIELDS

The kinematics of the gas in spiral disks are dominated by differential rotation and can be used to infer the distribution of mass (Section 4). The expected pattern of the velocity field for a well–ordered rotating disk is illustrated in Figure 12.1 of Giovanelli & Haynes (1988). Asymmetries and structure in global profiles can indicate that deviations from circular velocities in a single plane are present (Schulman *et al.* 1994). Maps of the velocity field are also useful for identifying warps, non–circular motions and disturbances due to bars, spiral arms or small companions.

Oval distortions in the inner regions of galaxies arise from gas streaming motions due to the non–symmetric potentials of bars. The iso–velocity contours near the centers are skewed, i.e. the kinematical major and minor axes are not perpendicular. Recent studies are those of NGC 1365 (van Moorsel *et al.* 1996) and NGC 1530 (Regan *et al.* 1996). The 21-cm line is not always the best spectral line for the study the dynamics of bars. In many cases the central regions of barred galaxies are devoid of H I (e.g. Ondrechen & van der Hulst 1989), and more information can be gained from CO mapping (e.g. Gerin *et al.* 1988; Sanqvist *et al.* 1988) and especially 2-D Fabry–Perot observations in the Hα line (e.g. Teuben *et al.* 1986). The disturbed velocity fields observed in barred galaxies can be modeled quite well by hydrodynamical gas flow models (Teuben *et al.* 1986), although several problems remain outstanding (cf. Sellwood 1991).

Streaming motions of gas across spiral arms are one of the best observational tests for the density–wave theory for spiral structure. For example Visser (1980a,b) showed that the velocity gradients observed by Rots (1975) across the arms of M81 are quite accurately modeled by the streaming of a cold, single–fluid gas which rotates in a potential having a spiral density–wave perturbation. Spiral streaming motions have been observed in several galaxies in H I , CO and Hα with amplitudes up to 90 km s^{-1} in the plane of the galaxy (Vogel *et al.* 1988; Ondrechen *et al.* 1989; Tilanus & Allen 1991; Canzian *et al.* 1993).

One of the — kinematically — most intriguing systems is the galaxy NGC 4826, nicknamed the "Evil Eye". Combined VLA B, C, and D array observations by Braun *et al.* (1992, 1994) revealed an regular H I gas disk extending to 14 kpc with a counter–rotating inner gas disk of 1 kpc radius. The amplitude and sense of rotation in the central regions are fully consistent with Hα and stellar absorption rotation curves (Rubin 1994). The velocity reversal is however only observed in the emission–line (H I and Hα) gas. The most likely explanation for this (at least for now) unusual system is the merger of two disk galaxies with anti–parallel spins. A similar velocity structure has been observed in the irregular galaxy NGC 4449. Its H I envelope extends to 14 times the optical size (Bajaja *et al.* 1994) and the large–scale velocity gradient along the major axis is opposite to that observed in H I and Hα over the optical image of the galaxy. This suggest that also NGC 4449 contains two distinct dynamical systems; detailed synthesis mapping of the inner regions is needed to distinguish between streaming motions along the main bar and counter–rotating gas disks as the cause of the observed velocity reversal.

4. H I as a Tracer of the Mass Distribution

H I rotation curves provide the most detailed quantitative evidence for mass discrepancies in galaxies. The distribution and kinematics of the H I is often sufficiently regular to allow an accurate determination of the circular velocities as a function of radius ("the rotation curve") out to the edges of the H I disk. In general, the coplanar circular motion of the cold gas in the outer regions is an excellent tracer of the radial force field and therefore of the distribution of mass. For most spirals the H I velocity field can be measured far beyond the edge of the visible components, in regions where no other observational technique can be used to obtain kinematical information. rotation curves derived from optical spectra (with high spatial resolution) can place the strongest constraints on the structure of the inner disk and bulge, while 21-cm line rotation curves likewise constrain the dark halo distribution.

4.1. ROTATION CURVES

Deriving a rotation curve from a two–dimensional velocity field is a complex process involving a number of assumptions. The most important one is the assumption that hydrogen clouds move in circular orbits around the galaxy center so that a gas disk can be modeled by a set of concentric rings each with its own orientation parameters and value for the circular velocity. All the parameters (rotation velocity, inclination and position angle of the line of nodes) can be determined by a least–squares fit to the observed velocity field. The inclinations used are derived directly from the velocity field and allow a first order correction for mildly warping gas disks for inclinations larger than about 50° (Begeman 1989). Large deviations from circular motions can be identified by comparing the observed velocity field with a model velocity field calculated from the adopted rotation curve (see also van der Kruit & Allen 1978).

In the beginning of the 1970's observations with large synthesis radio telescopes showed that H I rotation curves of spiral galaxies remain more or less flat even *beyond* the optical boundaries indicating that the integrated mass increases linearly with radius (since $\mathcal{M}_{\rm T}(R) \propto R\,V_{\rm cir}^2$) all the way to the last measured point. The extensive study of H I rotation curves by Bosma (1981a,b) showed the necessity of two–dimensional velocity information, since the gas in many galaxies showed disturbances due to bars, warps and in some cases due to interactions with other galaxies.

By combining the observed rotation curve with the known distribution of stars from the surface brightness profile and with the distribution of neutral gas, one can put constraints on the amount and radial distribution of luminous and dark matter. Mass models can be constructed by calculating the rotation curves for the gas disk, for the stellar bulge and disk component under the assumption of constant $\mathcal{M}/\mathcal{L}$ and subsequently fitting the (quadrature) sum of the curves to the observed rotation curve by varying the $\mathcal{M}/\mathcal{L}$ ratio of disk and bulge. In general the inner region of a galaxy rotation curve can be fitted very well, providing arguments for the so–called maximum disk hypothesis; however to fit the curves in the outer regions one has to assume the presence of an as yet invisible component. In his thesis Begeman (1987) showed for a sample of eight spiral galaxies that the amount of missing mass is about three times the amount of luminous mass. Simultaneously, Carignan and co–workers showed that rotation curves of low–luminosity disk galaxies tend to continue rising beyond their optical radius (e.g. Carignan & Freeman 1988; Jobin & Carignan 1990; Côté *et al.*

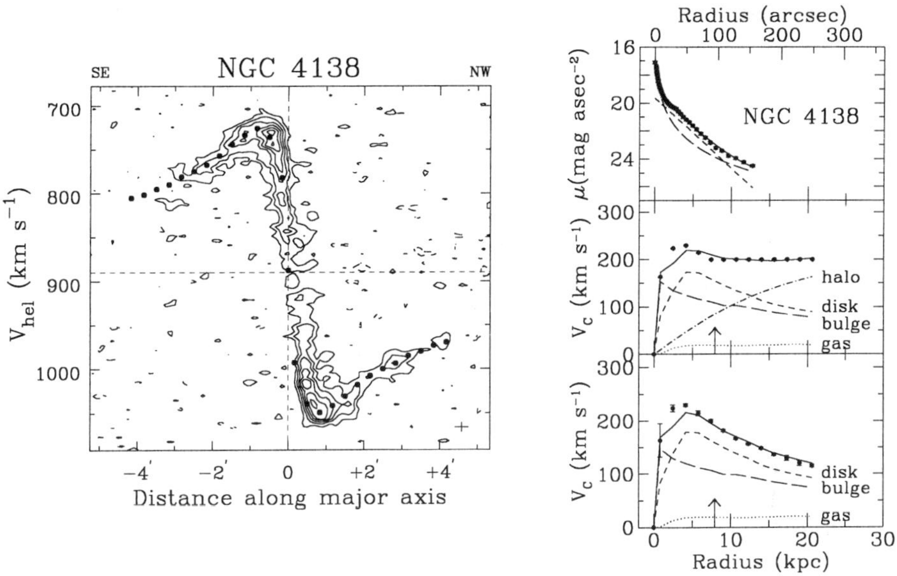

Figure 3. (left) Position–velocity map of the HI in NGC 4138 along the major axis. The strongly declining radial velocities in the outer regions are consistent with a declining rotation curve (lower panel on right–hand side), but also with flat rotation curve (middle panel) with a severely warped gas layer.
(right) The top panel shows the surface brightness profile; the lower panels shows the observed and model rotation curves for NGC 4138. Note that no dark halo is needed at all for the falling rotation curve model. From Broeils *et al.* (1995).

1991), and that the resulting dark–to–luminous mass ratios ($\mathcal{M}_{\rm dark}/\mathcal{M}_{\rm lum}$) are on average higher than for more luminous spirals. Casertano & van Gorkom (1991) reported on correlations between the slope of the rotation curve beyond the optical radius on the one hand and maximum rotation velocity, disk scale length and central surface brightness at the other. The picture that emerges is that of falling rotation curves for bright, compact galaxies, rising curves for low–luminosity galaxies and more–or–less flat rotation curves for the rest.

For a sample of 23 galaxies with extended HI rotation curves and CCD surface photometry Broeils (1992a, 1995) constructed mass models based on the maximum disk hypothesis. The maximum disk solution can be obtained nearly unambiguously for each galaxy and guarantees an upper limit for the disk $\mathcal{M}/\mathcal{L}$ and total luminous mass, and a lower limit for the dark mass and $\mathcal{M}_{\rm dark}/\mathcal{M}_{\rm lum}$ (inside the last measured point of the rotation curve). The models show that the dark matter fraction $\mathcal{M}_{\rm dark}/\mathcal{M}_{\rm lum}$ correlates with the maximum rotation velocity and spiral sub–type, and ranges from about 1 for Sab galaxies to nearly 10 for the extreme dwarf irregulars.

Based on optical studies, Kent (1988) has pointed out that about half of Sa galaxies show rotation curves that rise too slowly when compared with the light distribution and bulge dynamics, implying that the optical rotation curves are not measuring the circular velocity. In some of these objects, little or no dark matter is required. The optical studies however cannot place strong constraints on the existence of dark halos — or the lack thereof; however, few non–interacting Sa galaxies have been mapped in H I , primarily because their $\sigma_{\rm HI}$ is typically low (see Figure 1). Our ongoing H I and optical (emission and absorption line) study of a sample of H I –rich Sa galaxies indicates that these galaxies indeed contain relatively little non–luminous material. In this respect the results of VLA 21-cm observations of NGC 4138 are very interesting (Jore *et al.* 1995). The radial velocities of the H I gas beyond the optical disk drop by about 80 km s^{-1} , symmetrically with respect to the center (Figure 3). However, the observed velocity field is both consistent with a rotation curve with a nearly Keplerian decline and with a flat rotation curve in a strongly warped H I disk. The former model requires no dark matter at all; for the flat rotation curve a $\mathcal{M}_{\rm dark}/\mathcal{M}_{\rm lum}$ of 1.3 is found (Figure 3).

Another interesting result from the mass models is that the distribution of H I –to–dark mass ratios is much narrower than H I –to–stellar mass ratios, and that $\mathcal{M}_{\rm HI}/\mathcal{M}_{\rm dark}$ varies relatively little with morphological type. Even more intriguing is that in many galaxies the radial gas surface densities are very similar to the inferred dark halo surface densities (cf. Bosma 1981b; Carignan *et al.* 1990; Broeils 1992b). These two facts suggest that the distribution of neutral gas and dark matter are somehow related and perhaps coexisting in a flattened distribution. This hypothesis has lead to a number of proposals (Pfenniger *et al.* 1994; cf. Silk 1996) in which the bulk of the dark matter is in the form of cold gas, mostly in molecular form (see also chapter 1, p. 42). Future constraints on the flaring of the H I disk will tell whether the cold gas can be disk gas or has to form a more spherical distribution (see next section).

4.2. SHAPES OF DARK HALOS

A galaxy disk can be approximated by a self–gravitating, locally isothermal (i.e. the velocity dispersion of any of its components is not dependent on z) sheet. Then the total mass density can be expressed as

$$\rho(R,z) = \rho(R,0)\mathrm{sech}^2\ \frac{z}{z_0}$$

where the vertical scale parameter z_0 can vary with distance from the galactic center. Any layer that does not contribute significantly to the overall mass can be identified as a separate component, characterized by its own velocity dispersion σ_z and constrained by the mass distribution in the disk so that it fills a layer of full width at half peak value

$$z_{FWHP} = 1.7\,\sigma_z[2\pi G\,\rho(R,0)]^{-1/2}$$

following Shostak & van der Kruit (1984). See also Gilmore *et al.* (1989, Chapter 10).

The scale height of the H I gas in the Galaxy at $R = 25\,\mathrm{kpc}$ is about 10 times larger than it is at the solar circle (Burton 1988). The same kind of flaring behavior has been found in a number of external edge–on galaxies. Studies of face–on galaxies (summarized by Kamphuis 1993, p. 264; see also van der Hulst, 1996, 1997) have shown that the vertical velocity dispersion declines slowly with R from 10–13 km s^{-1} in the optical region to 6–8 km s^{-1} beyond R_{25}. The behavior of σ_z and z_{FWHM} with R can be used to estimate the vertical force field, and combining this with the radial distribution of mass from a rotation curve analysis can constrain the axis ratio of the dark halo distribution (Maloney 1993; Olling & van Gorkom 1995). Inside R_{25}, the stellar disk will dominate the potential and the z–thickness is expected to be rather small. Beyond R_{25}, the z–force field is the result of the contributions of the gas itself (self–gravity can not be ignored here!) and the dark halo. A more spherical halo will allow the H I layer to flare much more than a flattened halo. At present results are still inconclusive, but several ambitious studies are underway (Bosma 1991; Olling 1995, 1996a, 1996b).

It should be noted that until now the most significant results on shapes of dark halos have come from studies of polar ring galaxies. Polar rings (see also Section 5.2) are rings of gas, dust and stars orbiting at a large angle to the equator of a flattened galaxy, usually an S0. Measurements of both polar and equatorial rotation curves allows one to constrain both the radial and the vertical distribution of dark matter. Sackett & Sparke (1990; see also Sackett *et al.* 1994) show that for the proto-typical polar ring galaxy NGC 4650A, the dark halo is most likely flattened with a vertical-to-radial axis ratio of $0.3 \leq c/a \leq 0.4$. Unfortunately, polar ring galaxies are rather rare, so very few objects have been and can be studied in such detail.

5. Relationship of Σ_{HI} to Star Formation

The H I distribution outlines the location of diffuse gas. While molecular clouds are more directly linked to the star formation process, the H I nonetheless serves as a good tracer of the cool gas distribution and in the outer disk at least, a quantitative indicator of the gas surface density $\Sigma_{\mathbf{gas}}$. Considerations of the stability of an infinitely thin, uniformly rotating disk (cf. Binney & Tremaine 1987) lead to a criterion for disk stability, with separate criteria for gas and stars. This so–called Toomre stability criterion is close to the predicted gravitational instability threshold. Typically, this instability is evaluated as a threshold gas surface density Σ_c (Kennicutt 1989, chapter 7) above which collapse and subsequent star formation will occur

$$\Sigma_c = \alpha \frac{\kappa \sigma_v}{3.36G}, \quad \text{where} \quad \kappa = 1.41 \frac{V}{R}\left(1 + \frac{R}{V}\frac{dV}{dR}\right)^{1/2}$$

and $\alpha \sim 0.67$ for a thin gaseous disk (see Kennicutt 1989).

This type of analysis can be applied in many contexts. In the following section we explore several of the critical issues.

5.1. STAR FORMATION IN SPIRAL DISKS

The sequence of events that leads to the transition of diffuse H I clouds into the dense molecular clouds that eventually collapse to form stars is not known. In spiral disks, the molecular component as seen for example in the distribution of CO is more tightly associated with the dust clouds and star formation regions. Still in most galaxies, the H I follows well the optical and molecular spiral structure.

In the interacting M51 pair, the bright Sc NGC 5194 is a classic example of a "grand design" spiral galaxy whose prominence of spiral structure is likely driven by a tidal density wave. In other galaxies, spiral structure is weak and patchy ("flocculent") and density waves do not appear to play a significant role. The precise relationship of density waves to the triggering of star formation is not clear, but the comparison of the distribution and motions of the molecular, atomic and ionized components of the gas along with the H II regions, the radio continuum ridge and the magnetic field lines is beginning to unveil the details of the process in M51. The H I follows the spiral structure in the inner regions but, in the outer regions, it extends to large radii and appears to lie in a different plane (Rots *et al.* 1990). The velocity structure on the large scale is complex and likely results from the recent tidal interaction. On finer scales, the H I lies displaced from the non–thermal radio continuum and the dust lanes, suggesting that the offset H I results from the dissociation of molecular clouds by the star formation process (Tilanus & Allen 1991). As do the molecular and ionized components, the H I shows the streaming motions expected in nonlinear density wave models (Rand 1993).

While it is fairly clear that density waves can produce symmetric, high contrast spiral arms and that strong density waves can easily be generated by tidal encounters, one then wonders what causes the weaker, patchier flocculent spiral structure seen in the majority of spirals. Braine *et al.* (1993)

used observations of H I, Hα and ^{12}CO(1-0), ^{12}CO(2-1), ^{13}CO(1-0), ^{13}CO(2-1) to study the star formation process in the flocculent Sc galaxy NGC 4414. Despite the relatively modest rate of star formation, the neutral gas surface density in the disk is extremely high, reaching $\sim 50\,M_\odot\,pc^{-2}$, comparable to that seen in M51. Braine *et al.* hypothesize that the difference in the star formation rates of NGC 4414 and M51 may result because NGC 4414 is isolated. While the H I plays an important role, the molecular, radio continuum and Hα emissions are more critical to tracing the star formation process (Knapen *et al.* 1991). Further comparison of the H I, Hα and CO distributions and their respective ties to the star formation process are discussed in chapters 2 and 4 by Kenney and Rand.

There also exists an important category of luminous but low surface brightness (LSB) spirals that have been frequently overlooked in typical galaxy catalogs because of their LSB character. Detailed studies of the stellar populations of the luminous LSB's shows that they are blue and metal poor (McGaugh & Bothun 1994, de Blok *et al.* 1995). Indeed, their low metallicities suggest that they have formed stars very inefficiently over a Hubble time (McGaugh 1994). Van der Hulst *et al.* (1993) and de Blok *et al.* (1996) have shown that the luminous LSB galaxies have average σ_{HI} about two times lower than in HSB galaxies of the same type, so that the observed σ_{HI} indeed falls below the critical threshold for star formation. It hence appears that the giant H I disks in these luminous and typically isolated galaxies have evolved only very slowly. The importance of these objects in estimating the true density of galaxies in the universe has been emphasized by numerous authors (e.g. Disney 1976; Mo *et al.* 1994).

5.2. ORIGIN OF H I IN E'S AND S0'S

As mentioned in Section 2 and evident in Figure 1, the H I mass fraction and global surface density among the elliptical and S0 galaxies are generally lower than in later types. The distributions are also very broad, with some early types containing H I masses normal of Sb types or later and others none to stringent detection limits. It becomes curious why H I disks ever exist in such objects. Of fundamental issue is the possibility that the H I disks seen in the earliest types are the result of the recent accretion of gas–rich satellites or tidal material pulled from companions.

Elliptical galaxies are most likely triaxial stellar systems, with no preferential plane (disk). The observed flattening of the stellar distribution cannot be explained by rotation, since typical rotation velocities are measured to be ≤ 50 km s^{-1}. In contrast, the H I seen in the few ellipticals where it is found often lies in a rotating disk, though often found outside of the optical distribution and characterized by peculiar kinematics. Based on the statistical analysis of H I observations of some 150 E's, Knapp *et al.* (1985) concluded that the occurrence of H I in E's was fundamentally different from that seen in spirals in that there was no well defined "typical" H I content for E's, arguing for an external origin.

Perhaps the clearest example of a recent accretion event is seen in the dust and gas disk found in Centaurus A, NGC 5128. The main H I coincides with the prominent dust lane seen in optical images (van Gorkom *et al.* 1990), even to the extent of showing the same warped appearance. The disturbed kinematics observed in the outer portions of the dust disk suggest non–circular motions that are best interpreted by a capture scenario in which in the infalling material has not yet settled into stable orbits. Diffuse H I is also seen to follow the curvature of the outermost optical shells, with the H I displaced slightly to the outside of the optical ridge (Schiminovich *et al.* 1994). Another such example of H I associated with optical shells is the galaxy NGC 2865 (Schiminovich *et al.* 1995).

As in the E's, the H I content of S0's is highly varied. While about one fifth of the bright nearby S0's contain significant amounts of H I , many others that appear similar optically contain no detectable H I (Bregman *et al.* 1992). In S0 galaxies, the H I is often found in an annular distribution with the inner radius of the ring coincident with the outer edge of the optical disk. Based on a comparison of the observed H I distribution and kinematics of 24 early type galaxies, van Driel & van Woerden (1991) conclude that the S0's are most likely to have inner or outer H I rings, often tilted with respect to the stellar disk. In most of these rings, the azimuthally–averaged H I surface density is less than the critical threshold required for star formation. A particularly interesting case is the galaxy NGC 5102, optically an unusually blue peculiar S0 that shows the characteristics of a nuclear starburst with an age of a few $\times 10^8$ years. Hα emission outlines a conspicuous ring centered on the nucleus with a diameter of 1.7 kpc. van Woerden *et al.* (1994) find that the H I distribution is primarily concentrated in a 3.6 kpc wide ring with an average radius of 3.7 kpc (about $0.7 R_{25}$); within the Hα shell, H I is largely absent. The maximum azimuthally averaged H I surface density in the ring is $1.4\,\mathrm{M}_{\odot}\,\mathrm{pc}^{-2}$ comparable to that found in other S0's but the H I distribution resembles more nearly the distributions seen in early spirals. Based on the starburst characteristics, the distribution of the H I and the relatively high $\mathcal{M}_{\rm HI}/\mathcal{L}_{\rm B}$, van Woerden *et al.* propose that either this object has recently accreted a gas–rich dwarf or its gas disk has recently collapsed.

The occurrence of kinematically–decoupled gas and dust disks such as previously mentioned in Section 3.4 strongly argue for the recent external origin of this material. Appleton *et al.* (1990) have interpreted the asymmetric H I distribution and kinematics of the H I in NGC 5903 as a recent accretion event ($< 2\times10^8$ years ago) that has resulted in a gas disk oriented along the minor axis of the stellar system. Those authors also note that its close elliptical companion, NGC 5898 possesses a counterrotating Hα disk as reported by Bertola & Bettoni (1988). The polar rings mentioned in Section 4.2 also provide strong evidence for the formation of disks in S0's by the recent disruption and infall of gas–rich companions (Schweizer *et al.* 1989). Based on a new, largely successful survey for H I in E's, Huchtmeier (1994) concludes that in groups, gravitational interactions and mergers do indeed lead to the acquisition of H I by elliptical galaxies. However, while it is true that E and S0 galaxies in general inhabit the denser regions of the universe, it is not clear that the accretion or merger rate is high enough to explain all of the H I –rich early types nor the wide variation in properties observed among the S0 class (van Driel & van Woerden 1991).

5.3. OPTICALLY FAINT H I RICH GALAXIES

At the opposite end of the Hubble sequence, the H I contributes a significant fraction of the total galaxy mass, at least within the optical radius. The dwarf irregulars (dI's) are generally characterized by low luminosity, mass and surface brightness, and stellar populations that show bluer colors and metal abundances that are well below solar (Skillman *et al.* 1989). The simplest models of how a protogalaxy collapses to form a coherent object make two relevant predictions about the possible appearances of "young" galaxies in the local universe: (1) young stellar populations are characterized by low metallicities and (2) H I clouds, with densities just above the threshold density required for gravitational collapse, will evolve slowly. Using one or both of these clues, numerous studies have tried to identify and examine objects of suspected youth in order to probe not only the processes of galaxy and star formation but also to identify possible primordial (unprocessed) material in the local universe. In the dI's, which are often not rotation dominated, it is suggested that the star formation is depressed or suppressed by some combination of the low gas surface density and/or the consequences of low metal abundance on the formation of massive stars.

A first puzzle concerns how H I gas remains extended in systems that are not rotationally supported. Lo *et al.* (1993) have mapped the details of the H I kinematics and dynamics in nine low luminosity dI's including six members of the Local Group. Their H I masses are also very low, with $\mathcal{M}_{\rm HI}/\mathcal{L}_{\rm B}$ typical of normal dwarf galaxies. In only two of the galaxies is significant rotation detected; in all of them, random motions dominate. It is hard to understand why the H I gas has not already collapsed to form stars.

I Zw 18 is a double system of compact galaxies showing strong emission lines, embedded in an H I envelope that extends a factor of eight beyond the optical counterpart (Viallefond *et al.* 1987). Most importantly, its derived oxygen abundance is only 1/38th solar (Searle & Sargent 1972) making it the system of lowest known heavy element abundance at low redshift. Most recent evidence (Kunth *et al.* 1994) obtained from Hubble Space Telescope observations in the ultraviolet emphasizes the fact that while the heavy element abundance in I Zw 18 is low, it is not zero, and reinforces the importance of I Zw 18 to the determination of Big Bang nucleosynthesis constraints and the primordial He abundance.

I Zw 18, and other similar objects such as II Zw 40 (Brinks & Klein 1988) and ESO 400–G43 (Bergvall & Jörsäter 1988) represent the category of objects known as the blue compact dwarfs BCD's or H II galaxies, objects typically of low luminosity containing compact high surface brightness patches of current star formation. As in I Zw 18, metallicities are significantly sub–solar, placing stringent constraints on the number of starburst events, in some cases implying that stars are being formed for the first time. Several authors (e.g. Telles & Terlevich 1995) have argued that these objects are relatively isolated with respect to luminous galaxies so that their starburst events cannot have been triggered by recent interactions. H I synthesis observations of nine BCD's and their local environments by Taylor *et al.* (1993, 1994) have shown that four of the nine have H I companions with $\mathcal{M}_{\rm HI}$ between 2×10^7 to $5 \times 10^8 M_\odot$. These authors speculate that the starbursts have been triggered by the H I companions, but it is again not clear whether this is a universal mechanism responsible for the starbursts in H II galaxies.

Like I Zw 18, the object known as H I 1225+01 contains an extensive H I envelope dominated by two main clumps. Its H I –to–optical size ratio is the largest known, about 40 (Giovanelli *et al.* 1991). Optically, H I 1225+01 contains a small region of low optical surface brightness and ongoing star formation, referred to as the NE clump, that coincides with one of two peaks seen in the H I distribution. The extreme blue colors of the optical component imply that most of the light is produced by young stars, and the low metallicity derived for the H II regions implies that not more than two bursts, and perhaps only one, of star formation have taken place (Salzer *et al.* 1991). The second H I peak, called the SW clump, has no detected stellar emission to a B surface magnitude limit of 27 mag arcsec^{-2} (Salzer *et al.* 1991), giving it an inferred $\mathcal{M}_{\rm HI}/\mathcal{L}_{\rm B} > 200$. About 10% of the total gas is associated with a lesser peak between the two main ones (the "Bridge"), with a diffuse envelope of the system, and with very weak emission in the far periphery.

Recent VLA D–array maps of the H I emission with 40″ angular resolution, presented by Chengalur *et al.* (1995), show detected emission from all

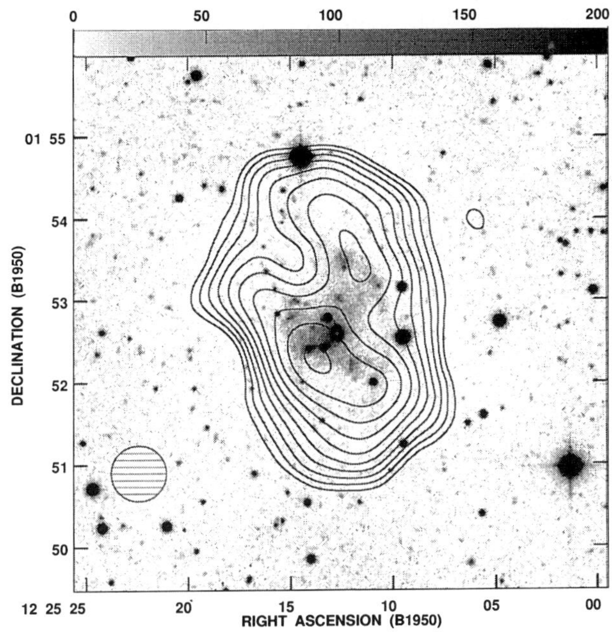

Figure 4. Overlay of the integrated H I emission (contours) on top of the optical emission (gray scales) for the northeastern component of H I 1225+01. The clean beam, $40'' \times 40''$ in size, is displayed in the bottom left corner. From Chengalur *et al.* (1995).

three components, although the SW clump also contains diffuse H I emission that is resolved out at the VLA. Figure 4 (from Chengalur *et al.*) shows contours of H I column density $N_{\rm HI}$ seen in the vicinity of the NE clump overlaid on a grey scale image of the optical emission detected by Salzer *et al.* The faint low surface brightness stellar emission corresponds quite closely to the highest H I column density contours. Star formation in the NE clump has occurred only above a (inclination corrected) threshold $N_{\rm HI}$ of $\sim 4.6 \times 10^{20}$ atoms cm^{-2}, as expected by the gravitational instability threshold criterion. Note that the H I peak is slightly displaced from the Hα peak.

Surprisingly, the SW clump has an observed peak $N_{\rm HI}$ that is higher than that of the NE clump, despite having an H I mass that is a nearly a factor of two smaller. This combined with its high axial ratio suggest that it is in fact a highly inclined disk, albeit one with a very small rotation velocity, ~ 13 km s^{-1} . If indeed edge–on, then the "face–on" H I column density is smaller than the gravitational instability threshold density, consistent with the lack of detectable stellar emission.

From the radial velocity field across H I 1225+01, it can be seen that the SW component is undergoing a "direct" encounter, consistent with a picture in which this component has been subject to considerable tidal damage by its companion. The Bridge is then material that has been pulled out from the SW clump and will presumably be captured by the NE clump. The spiral arm like features of the NE clump are also characteristic of two sided tidal responses. The morphology and kinematics of H I 1225+01 are thus qualitatively well explained as a consequence of tidal interaction between the main two components.

Both I Zw 18 and H I 1225+01 have large ratios of H I –to–optical diameters and low metallicities. Large extended H I disks appear to be a common characteristic of the most H I –rich dwarf galaxies, particularly isolated ones with unusually large ratios of $\mathcal{M}_{\rm HI}/\mathcal{L}_{\rm B}$ (van Zee *et al.* 1995). The interrelationship of the young stellar populations and the H I gas in such systems is relevant to an understanding of the low redshift Lyα absorbers and to the overall process of star formation during the epoch of galaxy formation as well as the present.

6. Environmental Effects

As we have seen in the preceding section, collisions, tides and accretion are often invoked to explain the peculiarities in the H I distributions of otherwise normal galaxies. It is well known that the most pathological morphologies such as seen in ring galaxies and the ultraluminous objects discovered by the *Infrared Astronomical Satellite* (IRAS) are linked to tides and collisions. Even in our local environment, the high velocity clouds manifest some disruptive phenomenon and the Magellanic Stream likely testifies of tidal disruption of the Milky Way–Magellanic Clouds system in the recent past (Lin *et al.* 1995; but see also Moore & Davis 1994).

Because the H I dominates in the outer disk, even beyond the optical edge, the H I distribution is an excellent tracer of disruptive events. Both collisions, tides and intracluster interactions are expected to remove gas from the fragile disk, lowering a disk galaxy's H I content or possibly creating a star–forming disk in a system previously without one. In the case of tidal interactions and gas sweeping, the H I distribution can also provide a glimpse of the disruptive event as the gas, removed from its normal disk location, is flung to large distances. Numerous examples exist of H I displaced from the parent galaxy by such processes. In this section, we discuss briefly several examples where the H I reveals the nature and consequences of the interaction of gas disks with their environment.

6.1. TIDES AND COLLISIONS

The history of tidal interaction in slow–moving encounters is vividly illustrated in the simple numerical simulations of Toomre & Toomre (1972). The resultant pathological optical morphologies are frequently reflected in disturbance of the H I disk, and the details of the H I distribution and velocity field relay the history of the encounter (e.g. Stanford & Balcells 1991; Smith 1994). In chapter 4 Rand discusses the disturbances in the disk and tidal spurs in the edge–on NGC 4631 (Rand 1994) in terms of tidal interactions with its companions. Likewise, the details of the interaction between M81 and M82 can been implied from the examination of H I streamers around M82 extending from the point where the disk flares (Yun *et al.* 1993).

Ring galaxies are believed to result from the direct collisions of a compact galaxy penetrating nearly perpendicular to the victim's disk. Hydrodynamic simulations of such encounters suggest that a large fraction of the interstellar gas winds up concentrated in the rings (Appleton & Struck–Marcell 1987) with star formation proceeding on the outer edge. In Arp 143, the H I lies in a crescent that lies slightly interior to the optical ring (Appleton *et al.* 1992). A crude CO(1–0) map made by Higdon *et al.* (1995) suggests that half of the molecular gas follows a similar distribution. In detail, the collisional model well predicts the location and kinematics of the gas and newly formed stars.

Although less clearly traceable, the link between extended H I distributions, star formation and the formation of disks has also been discussed in scenarios of severe tidal disruption and merging. Barnes & Hernquist (1992) have suggested that dwarf galaxies may form in the tidal debris. A likely candidate for such an event is the dwarf system located at the tip of one of the tidal tails seen in the Antennae NGC 4038/39 (Mirabel *et al.* 1992).

In compact groups, H I is often seen displaced from the member galaxies (Shostak *et al.* 1984; Williams & van Gorkom 1988), and its distribution relative to the member galaxies may serve as a good indicator of the evolutionary stage of the group itself (Williams *et al.* 1991). The large, quiescent H I cloud near the periphery of the cluster Abell 2634 contains two relatively small, actively star–forming galaxies, giving rise to the speculation that the cloud might be the remnants of a tidal or ram pressure event in the cluster environment (Giovanelli *et al.* 1995a). In all of these examples, the peculiarities of the H I distribution and velocity field signal the violence of such tides and mergers, and often provide the clues needed to constrain models of the recent events.

6.2. CLUSTERS

Dressler (1984) gives a detailed summary of evidence that galaxies in clusters are affected by their presence in such high density environments. Recent observations of the peculiar and disturbed morphological appearance of galaxies seen at high redshift by the Hubble Space Telescope reinforce the concept that rich clusters present a hostile environment to fragile spiral disks. Even in nearby clusters, evidence for on–going infall into cluster cores argues that collisional, tidal and stripping events continue to take place. Because of its association with the future star formation potential and its easy detectability, studies of the H I content and distribution within spiral disks in clusters has played a definitive role in our understanding of the morphology–altering processes in the nearby universe.

The first quantitative study of possible H I deficiency in cluster objects by Davies & Lewis (1973) targeted the Virgo cluster. Nearly immediately, Bottinelli & Gouguenheim (1974) pointed out the residual bias in H I content shown by Davies & Lewis: the most luminous galaxies appeared to be the most H I deficient. In the intervening years, many more clusters have been studied, and better comparison samples defining the standard of normal H I content have been assembled. Davies & Lewis' original conclusion that cluster galaxies tend to be H I deficient has been upheld.

The original controversy and concern about bias focussed on what measure of H I content (c.f. Section 2) is best suited for comparative studies. Because H I is a disk property whereas $\mathcal{L}_B$ contains contributions from both the disk and bulge, the globally averaged surface density σ_{HI} has been suggested as the best indicator. In practice the normalization for differences in scale uses the optical diameter, likewise a disk property, albeit producing a hybrid quantity. A common measure of H I deficiency compares the observed H I mass with that expected for a galaxy of a similar morphological type and linear diameter in the absence of any secular processes. Specifically, the H I deficiency parameter is defined as $< DEF > = log[\mathcal{M}_{HI}(T,D)]_o - log[\mathcal{M}_{HI}(T,D)_{obs}]$. Haynes *et al.* (1984) have determined the "normal" H I content through observations of a sample of isolated galaxies. Studies of the global H I content of spiral galaxies in clusters with different richness, evolutionary and X–ray properties allow the statistical evaluation of the degree of stripping and its cause. A recent review is given by Haynes (1990) based to a large extent on the compilations presented in Giovanelli & Haynes (1985) and Magri *et al.* (1988). Most authors conclude that spiral galaxies that pass within the central regions of rich clusters of galaxies that contain hot X–ray gas are stripped of their cool interstellar H I. Although the observed spirals retain some vestige of their spiral morphology, they are H I deficient by as much as a factor of ten relative to their more isolated counterparts. After disk fading, these stripped spirals will eventually appear as anemic spirals or even S0's, although this process is unlikely to be the sole cause of the predominance of early types in high density environments.

The exact mechanism by which the stripping occurs remains a matter of debate, partly because projection effects (i.e. only projected components of velocity and spatial separation are available) will tend to blur the analysis. Among the mechanisms most frequently discussed are galaxy–cluster tides, galaxy–galaxy collisions and galaxy–intracluster medium interactions, including ram pressure sweeping, thermal evaporation and turbulent viscous stripping. Different processes make somewhat different predictions; several clues are apparent:

1. The average H I deficiency is a strong function of distance from the cluster center.
2. The fraction of H I deficient galaxies increases with the cluster X–ray luminosity.
3. There is a one–to–one correspondence between H I deficiency and reduced H I size: the missing H I has been removed from the outermost regions of the galaxy.
4. The molecular component of H I deficient galaxies appears unaffected. Galaxies that are H I deficient by a factor of ten may be gas deficient only by a factor of three.
5. Early type galaxies are even more H I poor than late–type galaxies (Dressler 1986).
6. H I deficient spirals typically show redder colors and reduced star formation rates.

The understanding of the underlying cause of the H I deficiency requires details of the cluster structure and velocity field, the distribution and characteristics of the hot X–ray gas, and the relationship of the H I , molecular and stellar components of the disk. Because of its relative proximity, the Virgo cluster has been the target of many studies of environmental effects within the highest density regions because of the detail and variety of observational information available for it and its member galaxies. However, it should be noted that Virgo is not a particularly rich cluster and is only a moderate X–ray source. Dynamically, it consists of two distinct populations: the ellipticals forming a relaxed, collapsed core, while the spirals are found in clouds or groups, some of which are clearly infalling.

The details of the H I distribution in 17 bright Virgo spirals have been examined by Cayatte *et al.* (1994). Making use of the radial H I surface density distributions $\sigma_{\rm HI}(R)$, those authors are able to investigate both the H I H I deficiency, H I extent and overall H I surface density of the Virgo spirals. They find that all mechanisms — ram pressure, viscous stripping and tides — are likely to play a role, and their dominance can be distinguished by examining the details of the H I distribution in each galaxy. The galaxies with smallest H I sizes have relatively normal H I surface densities, as expected for galaxies which are swept by ram pressure interaction with the hot intracluster gas. Turbulent viscous sweeping predicts an overall reduction in the disk σ_{HI}, as seen in deficient galaxies with only slightly shrunken H I disks. Other galaxies have some disturbance of their H I appearance which implies distant tides at work.

The complexity of cluster structure and evolutionary state blur the picture. H I synthesis maps are also available for a sample of galaxies in the Hydra cluster, Abell 1060, and, while the cluster is similar to Virgo in many respects, the Hydra spirals do not appear to be either H I deficient or H I small. At least some of the normal spirals in the Hydra direction may actually be members of a foreground group (McMahon *et al.* 1992). In Abell 1367, there are several peculiar blue spirals found near the edge of X–ray emission in which the H I is displaced from the optical disk, and enhanced radio continuum emission suggests a higher than normal star formation rate (Dickey & Gavazzi 1991). This phenomenon suggested to Gavazzi & Jaffe (1986) that the interaction with the hot X–ray gas causes compression of the interstellar clouds and actually induces star formation on the forward side of an infalling galaxy further consuming the H I gas. Sometime in the future, these objects will appear to be strongly H I deficient, and after significant disk fading, red.

As described by Gunn & Gott (1972), the ram pressure sweeping model predicts the removal of H I from the outer disks of spiral galaxies traveling at high velocity through the hot intracluster medium. In this model, it is expected that stripping will be most effective in low mass systems, whose lowered gravitational restoring forces make them even more vulnerable. The destruction of a low mass galaxy by tides with other galaxies or the cluster potential itself may be witnessed in the presence of a small H I cloud halfway between NGC 4472 and the gas–deficient dI UGC 7636 (Henning *et al.* 1993). However as noted by Hoffman *et al.* (1988), the dwarfs in Virgo are not significantly more H I deficient than the massive spirals, a contradiction of the ram pressure concept.

7. Future prospects

Throughout this review, we have tried to point out the many issues relating to the structure, kinematics and dynamics of galaxies that H I studies address. The last two decades have brought to maturity the techniques of aperture synthesis mapping with a corresponding wealth of detail. Broad–band spectrometers and low noise amplifiers allow sensitive searches for

H I over wide ranges of redshift space. The future holds equal potential, as instruments as constructed and upgraded. Within the next few years, new telescopes now under construction will be operational. The Giant Meterwave Radio Telescope GMRT near Poona in India has capabilities to map redshifted H I in galaxies and clusters and to search for protogalaxies and other early hierarchical structures. The Westerbork Synthesis Radio Telescope WSRT has new front-ends covering 260-450 MHz and 700-1200 MHz, also suited for observing H I over a large range of redshifts. The 100–m Green Bank Telescope GBT is a fully steerable, unblocked aperture antenna. Its "clean" beam will allow the study of faint structures in the Galactic H I , as well as sensitive searches in redshift space. In the same interval, the upgrade of the Arecibo 305–m antenna to a Gregorian optical system will give both lower noise and greater frequency flexibility, permitting incomparable sensitivity for objects in the Arecibo declination range. Future advances in digital technology will provide significant new capabilities. Digital backends will provide broader bandwidths and more channels for both single dish and aperture synthesis applications. Both hardware and software will be required to respond to the ever–deteriorating radio frequency interference environment through pre–detection filtering and excision techniques. A good idea of the prospects of these capabilities is given by several contributions in Bremer *et al.* (1996)

In the future, the routine acquisition of aperture synthesis maps will allow statistical studies, nowadays limited by the relatively small number of galaxies mapped with adequate sensitivity and resolution. Already underway is the WHISP (Westerbork H I Survey of Spiral Galaxies), a survey of the H I component in spiral galaxies with the WSRT. Its aim is to obtain maps of the distribution and velocity structure of H I in 500 to 1000 nearby galaxies. The resultant maps will serve as a basis for research in many areas, for example: dark matter halos, the effects of environment on the structure and sizes of H I disks, intergalactic H I and the cosmological distance scale through the Tully–Fisher relation.

Although it lies somewhat peripheral to the current discussion and is nearly absent from this review, the use of the Tully–Fisher relation will continue as a powerful tool for the measurement of peculiar motions in the nearby universe. The detailed mapping of the H I distribution in spiral disks is key to an understanding of the physical basis for the Tully–Fisher relation and a determination of its intrinsic scatter. While the extension to higher redshift will involve both the H I Tully–Fisher relation and its optical analog (Vogt *et al.* 1993), the H I method appears to be tighter than the optical one (Giovanelli *et al.* 1997a, 1997b). Future H I applications will provide a greater sampling of the local peculiar velocity field, important in order to reduce the volume averaging in reconstructions of the local density field.

While at first glance, the cool H I disks in galaxies may seem rather mundane, indeed, the wealth of detail provided by H I observations reveals a picture of great complexity and provides the clues to innumerable puzzles in cosmology. Undoubtedly, future studies of the H I in galaxies will both provide new understanding and raise new questions.

Acknowledgements

The National Astronomy and Ionosphere Center is operated by Cornell University under a cooperative agreement with the National Science Foundation. This work has been supported by NSF grants AST–9023450 and AST–9218038.

References

Appleton, P.N., Pedlar, A. and Wilkinson, A. 1990, *Ap.J.* , 357, 426.
Appleton, P.N., and Struck–Marcell, C. 1987, *Ap.J.* , 318, 103.
Appleton, P.N., Schombert, J.M. and Robson, E.I. 1992, *Ap.J.* , 385, 491.
Bahcall, J.N., Jannuzi, B.T., Schneider, D.P., Hartig, G.F., Bohlin, R., and Junkkarinen, V. 1991, *Ap.J.(Lett.)* 377, L5.
Bajaja, E., Huchtmeier, W.K. and Klein, U. 1994, *Ast.Ap.* , 285, 385.
Barnes, J.E. and Hernquist, L. 1992, *Nature* , 360, 715.
Begeman, K.G. 1987, Ph.D. thesis, University of Groningen.
Begeman, K.G. 1989, *Ast.Ap.* , 223, 47.
Bergvall, N. and Jörsäter, S. 1988, *Nature* , 331, 589.
Bertola, F. and Bettoni, D. 1988, *Ap.J.* , 329, 102.
Binney, J. 1992, *Ann.Rev.Ast.Ap.* 30, 51.
Binney, J. and Tremaine, S. 1987, *Galactic Dynamics*, (Princeton University Press: Princeton), pp. 362.
Bosma, A. 1981a, *A.J.* 86, 1791.
Bosma, A. 1981b, *A.J.* 86, 1825.
Bosma, A. 1991, in *Warped disks and inclined rings around galaxies*, eds. S. Casertano, P. Sackett, F. Briggs (Cambridge University Press: New York), p., 181.
Bottema, R., Shostak, G.S., and van der Kruit, P.C. 1987, *Nature* 328, 401.
Bottema, R. 1995, *Ast.Ap.* , 295, 605.
Bottema, R. 1996, *Ast.Ap.* , 306, 345.
Bottinelli, L. and Gouguenheim, L. 1974, *Ast.Ap.* , 36, 461.
Braine, J., Combes, E. and van Driel, W. 1993, *Ast.Ap.* , 280, 451.
Braun, R, Walterbos, R.A.M. and Kennicutt, R.C. 1992, *Nature* , 360, 442.
Braun, R. and Walterbos, R.A.M. 1992, *Ap.J.* , 386, 120.
Braun, R., Walterbos, R.A.M., Kennicutt, R.C. and Tacconi, L.J. 1994, *Ap.J.* , 420, 558.
Braun, R. 1995, *Ast.Ap.Suppl.* , 114, 409.
Braun, R. 1997, *Ap.J.* , in press.
Bregman, J.N., Hogg, D.E. and Roberts, M.S. 1992, *Ap.J.* , 387, 484.
Bremer, M.N., van der Werf, P.P., Röttgering, H.J.A. and Carille, C.L. 1996, *Cold Gas at High Redshift*, Kluwer Acad. Publ. ASSL Vol. 206.
Briggs, F.H. 1990, *Ap.J.* , 352, 15.
Brinks, E. and Klein, U., 1988, *M.N.R.A.S.* , 231, 63.
Broeils, A.H. 1992a, Ph.D. thesis, University of Groningen.
Broeils, A.H. 1992b, *Ast.Ap.* , 256, 19.

Broeils, A.H. and van Woerden, H. 1994, *Ast.Ap.Suppl.* , 107, 129.

Broeils, A.H. 1995, in *"Dark Matter"*, AIP Conference Proceedings #336, eds. Stephen S. Holt and Charles L. Bennet, p. 125.

Burton, W.B. 1988, in *Galactic and Extragalactic Radio Astronomy*, eds. G.L. Verschuur and K.I. Kellermann, (Springer–Verlag: New York), pp. 295.

Burton, W.B. and te Linkel Hekkert, P. 1986, *Ast.Ap.Suppl.* , 65, 427.

Canzian, B., Allen, R.J. and Tilanus, R.P.J. 1993, *Ap.J.* , 406, 457.

Carignan, C. and Freeman, K.C. 1988, *Ap.J.(Lett.)* , 332, L33.

Carignan, C., Charbonneau, P., Boulanger, F. and Viallefond, F. 1990, *Ap.J.* , 234, 43.

Carilli, C.L. and van Gorkom, J.H. 1992, *Ap.J.* , 399, 373.

Casertano, S. and van Gorkom, J.H. 1991, *A.J.* , 101, 1231.

Cayatte, V., Kotanyi, C., Balkowski, C. and van Gorkom, J.H. 1994, *A.J.* , 107, 1003.

Chengalur, J.N., Giovanelli, R. and Haynes, M.P. 1995, *A.J.* , 109, 2414.

Corbelli, E., Schneider, S.E. and Salpeter, E.E. 1989, *A.J.* , 97, 390.

Corbelli, E. and Salpeter, E.E. 1993, *Ap.J.* , 419, 104.

Côté, S., Carignan, C. and Sancisi, R. 1991, *A.J.* , 102, 904.

Davies, R.D. and Lewis, B.M. 1973, *M.N.R.A.S.* , 165, 231.

de Blok, W.J.G., van der Hulst J.M. and Bothun, G.D. 1995, *M.N.R.A.S.* 274, 235.

de Blok, W.J.G., McGaugh, S.S. and van der Hulst J.M. 1996, *M.N.R.A.S.* , 283, 18.

de Vaucouleurs, A., de Vaucouleurs, G., Buta, R.J., Corwin, H.G., Jr., Paturel, G. and Fouqué, P. 1991, *Third Reference Catalogue of Bright Galaxies* (Springer–Verlag: New York).

Dickey, J.M. and Gavazzi, G. 1991, *Ap.J.* , 373, 347.

Dickey, J.M and Lockman, F.J. 1990, *Ann.Rev.Ast.Ap.* , 28, 215.

Dickey, J.M., Brinks, E. and Puche, D. 1992, *Ap.J.* , 385, 501.

Dickey, J.M. and Brinks, E. 1993, *Ap.J.* , 405, 153.

Disney, M. 1976, *Nature* , 263, 573.

Dressler, A. 1984, *Ann.Rev.Ast.Ap.* , 22, 185.

Dressler, A. 1986, *Ap.J.* , 301, 55.

Gavazzi, G. and Jaffe, W. 1986, *Ap.J.* , 310, 53.

Gerin, M., Nakai, N. and Combes, F. 1988, *Ast.Ap.* , 203, 44.

Gilmore, G., King, I.R. and van der Kruit, P.C. 1989, *The Milky Way as a Galaxy*, (University Science Books: Mill Valley, CA.), pp. 206–211.

Giovanelli, R. and Haynes, M.P. 1985, *Ap.J.* , 292, 404.

Giovanelli, R. and Haynes, M.P. 1988 in *Galactic and Extragalactic Radio Astronomy*, eds. G.L. Verschuur and K.I. Kellermann, (Springer-Verlag: New York), pp. 522.

Giovanelli, R. and Haynes, M.P. 1990, *Ann.Rev.Ast.Ap.* , 29, 499.

Giovanelli, R., Williams, J. and Haynes, M.P. 1991, *A.J.* , 101, 1242.

Giovanelli, R., Scodeggio, M., Solanes, J.–M., Haynes, M.P., Arce, H. and Sakai, S. 1995a, *A.J.* , 109, 1451.

Giovanelli, R., Haynes, M.P., Herter, T., Vogt, N.P., Wegner, G., Salzer, J.J., da Costa, L.N., and Freudling, W., 1997a, *A.J.* , 113, 22.

Giovanelli, R., Haynes, M.P., Herter, T., Vogt, N.P., da Costa, L.N., Freudling, W., and Salzer, J.J. 1997b, *A.J.* , 113, 53.

Gunn, J.E. and Gott, J.R. 1972, *Ap.J.* , 176, 1.

Haynes, M.P. 1990, in *Clusters of Galaxies*, eds. W.R. Oegerle, M.J. Fitchett and L. Danly, (Cambridge University Press: New York), pp. 177.

Haynes, M.P., Giovanelli, R. and Chincarini, G.L. 1984, *Ann.Rev.Ast.Ap.* ,, 22, 445.

Henning, P., Sancisi, R. and McNamara, B. 1993, *Ast.Ap.* , 268, 536.

Higdon, J.L., Smith, B.J., Lord, S.D. and Rand, R.J. 1995, *Ap.J.(Lett.)* , 438, L79.

Hoffman, G.L., Helou, G. and Salpeter, E.E. 1988, *Ap.J.* , 324, 75.

Huchtmeier, W.K. 1994, *Ast.Ap.* , 286, 389.

Jobin, M. and Carignan, C. 1990, *A.J.* , 100, 648.

Jore, K.P., Broeils, A.H. and Haynes, M.P. 1996, *A.J.* 112, 438.

Kamphuis, J.J. 1993, Ph.D. thesis, University of Groningen.

Kennicutt, R.C. 1989, *Ap.J.* , 344, 685.
Kent, S.M. 1988, *A.J.* , 96, 514.
Knapen, J.H., Beckman, J.E., Cepa, J., van der Hulst, J.M. and Rand, R.J. 1991, *Ap.J.(Lett.)* , 385, L37.
Knapp, G.R., Turner, E.L. and Cunniffe, P.E. 1985, *A.J.* , 90, 454.
Kulkarni, S.R. and Heiles, C. 1988 in *Galactic and Extragalactic Radio Astronomy*, eds. G.L. Verschuur and K.I. Kellermann, (Springer-Verlag: New York), Chapter, 3.
Kunth, D., Lequeux, J., Sargent, W.L.W. and Viallefond, F. 1994, *Ast.Ap.* , 282, 709.
Lin, D.N.C., Jones, B.F. and Klemola, A.R. 1995, *Ap.J.* , 439, 652.
Lo, K.-Y., Sargent, W.L.W. and Young, L. 1993, *A.J.* , 106, 507.
Lockman, F.J. 1983, *Ap.J.* , 283, 90.
Magri, C., Haynes, M.P., Forman, W., Jones, C. and Giovanelli, R. 1988, *Ap.J.* , 333, 136.
Maloney, P. 1993, *Ap.J.* 414, 41.
McGaugh, S. 1994, *Ap.J.* , 426, 135.
McGaugh, S. and Bothun, G.D. 1994, *A.J.* , 107, 530.
McMahon, P., Richter, O.-G., van Gorkom, J.H. and Ferguson, H.C. 1992, *A.J.* , 103, 399.
Mirabel, I.F., Dottori, H. and Lutz, D. 1992, *Ast.Ap.* 256, 19.
Mo., H.J., McGaugh, S.S. and Bothun, G.D. 1994, *M.N.R.A.S.* , 267, 129.
Moore, B. and Davis, M. 1994, *M.N.R.A.S.* 270, 209.
Morris, S.L., Weymann, R.J., Savage, B.D. and Gilliland, R.L. 1991, *Ap.J.(Lett.)* , 377, L21.
Olling, R.P. 1995, *A.J.* , 110, 591.
Olling, R.P. 1996a, *A.J.* , 112, 457.
Olling, R.P. 1996b, *A.J.* , 112, 481.
Olling, R.P. and van Gorkom, J.H. 1995, in *"Dark Matter"*, AIP Conference Proceedings #336, eds. Stephen S. Holt and Charles L. Bennet, p. 121.
Ondrechen, M.P. and van der Hulst, J.M. 1989, *Ap.J.* 342, 29.
Ondrechen, M.P., van der Hulst, J.M. and Hummel, E. 1989, *Ap.J.* 342, 39.
Pfenniger, D., Combes, F. and Martinet, L. 1994, *Ast.Ap.* 285, 79.
Rand, R.J. 1993, *Ap.J.* , 410, 68.
Rand, R.J. 1994, *Ast.Ap.* , 285, 833.
Regan, M.W., Teuben, P.J., Vogel, S.M. and van der Hulst, J.M., 1996, *A.J.* 112, 2549.
Roberts, M.S. 1975, in *Galaxies and the Universe*, eds. A. Sandage, M. Sandage and J. Kristian, (University of Chicago Press: Chicago), pp. 309.
Roberts, M.S. and Haynes, M.P. 1994, *Ann.Rev.Ast.Ap.* , 32, 115 (RH94).
Rots, A.H., Bosma, A., van der Hulst, J.M., Athanassoula, E. and Crane, P.C., 1990, *A.J.* , 100, 387.
Rots, A.H. 1975, *Ast.Ap.* , 45, 43
Rowan-Robinson, M. 1985, *The Cosmological Distance Ladder*, (W.H. Freeman: New York), pp 184-191.
Rubin, V.C. 1994, *A.J.* , 107, 173.
Sackett, P.D. and Sparke, L.S. 1990, *Ap.J.* 361, 408.
Sackett, P.D., Rix, H.-W., Jarvis, B.J. and Freeman, K.C. 1994, *Ap.J.* , 436, 629.
Salpeter, E.E. 1995, in *The Physics of the Interstellar and Intergalactic Medium*, eds. A. Ferrara, C.F. McKee, C. Heiles, & P.R. Shapiro, (PASP: San Francisco), in press.
Salzer J.J., di Serego Alighieri, S., Matteucci, F., Giovanelli, R. and Haynes, M.P. 1991, *A.J.* , 101, 1258.
Sancisi, R. 1980, in *The Structure and Evolution of Nearby Galaxies*, eds. S.M. Fall and D. Lynden-Bell. (Cambridge Univ. Press: New York), pp. 149.
Sancisi, R. and Allen, R.J. 1979, *Ast.Ap.* 74, 73.
Sanqvist, Aa., Elfhag, T and Jörsäter, S. 1988, *Ast.Ap.* , 201, 223.
Schiminovich, D., van Gorkom, J.H., van der Hulst, J.M. and Kasow, S. 1994, *Ap.J.(Lett.)* , 423, L101.

Schulman, E., Bregman, J. and Roberts, M.S. 1994, *Ap.J.* , 423, 180.
Schweizer, F., van Gorkom, J.H. and Seitzer, P. 1989, *Ap.J.* , 338, 770.
Searle, L. and Sargent, W.L.W., 1972 *Ap.J.* , 173, 25.
Selwood, J.A. 1991, in *Evolution of Interstellar Matter and Dynamics of Galaxies*, eds. J. Palou
Shostak, G.S. and van der Kruit P.C. 1984, *Ast.Ap.* , 132, 20.
Shostak, G.S., Sullivan, W.T. III and Allen, R.J. 1984, *Ast.Ap.* , 139, 15.
Silk, J. 1996, in the proceedings of the 1993 Les Houches Summer School on Theoretical Physics on *Cosmology and Large-Scale Structure*, ed. R. Schaeffer *et al.* , Elsevier, p. 75.
Skillman, E.D., Kennicutt, R.C. and Hodge, P.W. 1989, *Ap.J.* , 347, 875.
Smith, B.J. 1994, *A.J.* , 107, 1695.
Sparke, L.S. and Casertano, S. 1988, *M.N.R.A.S.* 234, 873
Stanford, S.A. and Balcells, M. 1991, *Ap.J.* , 370, 118.
Taylor, C., Brinks, E. and Skillman, E. 1993, *A.J.* , 105, 128.
Taylor, C., Brinks, E., Pogge, R.W. and Skillman, E. 1994, *A.J.* , 107, 971.
Telles, E. and Terlevich, R. 1995, *M.N.R.A.S.* , 275, 1.
Teuben, P.J., Sanders, R.H., Atherton, P.D and van Albada, G.D. 1986, *M.N.R.A.S.* 221, 1.
Tilanus, R.P.J. and Allen, R.J. 1991, *Ast.Ap.* , 244, 8.
Toomre, A. and Toomre, J. 1972, *Ap.J.* , 178, 623.
van Driel, W. and van Woerden, H. 1991, *Ast.Ap.* , 243, 71.
van Gorkom, J.H., van der Hulst, J.M., Haschick, A.D. and Tubbs, A.D. 1990, *A.J.* , 99, 1781.
van Gorkom, J.H. 1992, in *The Environment and Evolution of Galaxies*, eds. J.M. Shull, H.A. Thronson, Jr. (Kluwer Academic Publishers: New York), p., 345.
van Gorkom, J.H., van Albada, T.S., Cornwell, T.J. and Sancisi, R. 1993, in preparation.
van der Hulst, J.M., Skillman, E.D., Smith, T.R., Bothun, G.D., McGaugh, S.S. and de Blok, W.J.G. 1993, *A.J.* , 106, 548.
van der Hulst, J.M. 1996, in *High Sensitivity Radio Astronomy* ed. N. Jackson and R.J. Davis, Cambridge Univ. Press, p.106
van der Hulst, J.M. 1996, in *Minnesota lecture series on HI in the Universe*, ed. E.D. Skillman, ASP. Conf Series. p.47
van der Hulst, J.M. 1997, in *The Physics of Galactic Halos* ed. H. Lesch, R-J. Dettmar, U Mebold and R. Schlickeiser, Akademie Verlag, p. 41
van der Kruit, P.C. and Allen, R.J. 1978, *Ann.Rev.Ast.Ap.* 16, 103.
van der Kruit, P.C. 1981, *Ast.Ap.* , 99, 298.
van Moorsel, G.A., Jörsäter, S. and Lindblad, P.O., 1996 *A.J.* , 00, 000.
van Woerden, H., van Driel, W., Braun, R. and Rots, A.H. 1994, *Ast.Ap.* , 269, 15.
van Zee, L., Haynes, M.P. and Giovanelli, R. 1995, *A.J.* , 109, 990.
Viallefond, F., Lequeux, J., Comte, G., 1987, in *Starbursts and Galaxy Evolution*, ed. T.X. Thuan, T. Montmerle, and J. Tran Thanh Van (Editions Frontières: Gif Sur Yvette), 139.
Visser, H.C.D. 1980a, *Ast.Ap.* , 88, 149.
Visser, H.C.D. 1980b, *Ast.Ap.* , 88, 159.
Vogel, S.N., Kulkarni, S.R. and Scoville, N.Z. 1988, *Nature* , 334, 402.
Vogt, N.P., Herter, T., Haynes, M.P. and Courteau, S. 1993, *Ap.J.(Lett.)* 415, L95.
Williams, B.A. and van Gorkom, J.H. 1988, *A.J.* , 95, 352.
Williams, B.A, McMahon, P.M. and van Gorkom, J.H. 1991, *A.J.* , 101, 1957.
Yun, M.S., Ho, P.T.P. and Lo, K.Y. 1993, *Ap.J.(Lett.)* 411, L17.

THE WARM IONIZED MEDIUM AND THE DISK-HALO CONNECTIC IN GALAXIES

RICHARD J. RAND
Dept of Physics and Astronomy
Univ. of New Mexico
800 Yale Blvd, NE
Albuquerque, NM 87131
U.S.A.

1. Introduction

The Warm Ionized Medium (WIM) or Diffuse Ionized Gas (DIG) is intimately related to the basic structure of the ISM as a whole and the dynamical processes which shape it, both in the Milky Way and external galaxies. For instance, the distribution of H I layers, radio and X-ray halos, the rate and distribution of star formation, the raising of gas out of the plane by stellar winds and ixsupernovae, and ways of creating holes in the ISM are all issues on which the study of the WIM has bearing. Although this phase has been recognized in the Milky Way ISM since 1971 (Reynolds *et al.* 1971), it is only within the last five years that we have begun to study the WIM in external galaxies in earnest.

This review will first summarize our observational understanding of the WIM of the Milky Way, also known as the Reynolds layer. Several theories of the origin and maintenance of the WIM will then be presented. These should provide ideas to keep in mind when we turn to observations of external galaxies – the main topic of this paper. A recent but more extensive review of the WIM in galaxies can be found in Dettmar (1992).

2. The WIM of the Milky Way

The local WIM of the Milky Way has been studied extensively by Reynolds and collaborators, and is often referred to as the Reynolds layer, after his report of the discovery of the diffuse Hα background in 1971 (see Reynolds

J. M. van der Hulst (ed.), The Interstellar Medium in Galaxies, 105–128.

1993 and references therein). It is studied primarily through its diffuse emission in Hα and other optical lines, and through pulsar dispersion measures (DM). Extinction limits the observable Hα emission to within 2–3 kpc of the Sun. Heiles (1994) reports radio recombination line emission deeper into the inner Galaxy from H I "worms" – vertical filaments of gas which reach a few hundred pc off the plane – and identifies this ionized gas as the inward, more powerful extension of the WIM.

The local WIM occupies a significant fraction of the interstellar volume – about 20% – and Kulkarni & Heiles (1988) present evidence that the filling factor may rise to > 0.4 at 1 kpc above the midplane. The mean density in the midplane is about 0.03 cm^{-3}, while the scale-height is about 900 pc. Thus, the WIM is more extended in the z direction than the z-extended H I component known as the Lockman layer, which has a scale height of 300 pc. The temperature of the WIM is estimated from the widths of optical emission lines to be about 8000 K. Since the exponential scale-height for gas at this temperature is only 60 pc (Dettmar 1992), there must be another form of support, presumably in ordered or turbulent motions. The surface density of the WIM in the Solar Neighborhood is 1.6 $M_{\odot}$pc^{-2}, or about one third of the local H I surface density. Hydrogen in the WIM is almost completely ionized (see below). Fabry-Perot images of the diffuse Hα emission reveal that the WIM is present in every direction, and has a patchy, filamentary structure on all observed angular scales ($> 1^{\circ}$), superposed on a more diffuse background. We do not yet have a fully sampled Hα survey of the WIM, but such a survey is planned by R. Reynolds to begin in 1996 with a Fabry-Perot instrument known as WHAM (Wisconsin Hα mapper, see Reynolds, 1977), with an angular resolution of 1°and a velocity resolution of 12 km s^{-1} . Other optical lines will be studied as well.

To keep the WIM ionized, an ionizing power of at least 1.0×10^{-4} erg s^{-1} per cm^{-2} of disk is required. Of the possible sources of ionization, only O stars comfortably exceed this requirement. While maintenance of the WIM requires only 15% of the O-star Lyman continuum production rate, it would require essentially 100% of the kinetic energy provided by supernovae. Hence, O stars are favored, but it is not clear that the ISM contains paths of sufficiently low H I column density to avoid absorption of the ionizing photons before they can travel hundreds of pc or more from the stars (the so-called porosity problem). For instance, in the ionization models of Sokolowski (1994), an H I column of $\gtrsim 10^{18}$ cm^{-2} is sufficient to prevent complete ionization of the local WIM.

Ratios of forbidden optical emission lines to Hα and Hβ distinguish the WIM from H II regions (see Table 1). The reader is cautioned, though, that these forbidden lines have been observed only at low latitudes, and could differ in the high-z gas. Also shown in Table 1 are the observed ratios in the diffuse gas in NGC 891, which we will return to later. Such enhanced ratios are also found in the interstellar "froth" in irregular galaxies (Hunter & Gallagher 1990), and this gas is also believed to be largely photoionized. The limits on [O I] imply that H is almost completely ionized, since the ionization of O is closely tied to that of H by a charge-exchange reaction. The enhanced [S II] and diminished [O III] emission are expected for gas ionized by a "dilute" radiation field, i.e. one with a low Ionization Parameter, U. This is defined as

$$U = \frac{L}{4\pi n c R_s^2} \propto (L f^2 n)^{1/3} \tag{1}$$

where L is the ionizing photon luminosity, R_s is the Strömgren radius of the ionized region, n is the local density and f is the filling factor. Hence, U represents the balance between ionizations and recombinations.

TABLE 1. Emission Line Ratios

	NGC 891 $R = 3$ kpc $z = 0.5$ kpc	NGC 891 $R = 3$ kpc $z = 1$ kpc	Reynolds layer	HII regions
[N II]/Hα	0.6	1.1	0.3	0.3
[S II]/Hα	0.3	0.6	0.3	0.1
[O III]/Hα	<0.15	-	0.06	0.5
[O I]/Hα	<0.05	-	<0.02	0.003

Photoionization models which reproduce the line ratios for the Milky Way WIM were developed in a classic paper by Mathis (1986), and the issue has been recently readdressed by Domgörgen & Mathis (1994) and Sokolowski (1994). However, a new upper limit by Reynolds & Tufte (1994) on He I 5876 Å recombination emission in two directions through the WIM poses a challenge to photoionization models by implying that He^+/H^+ is at most 0.03. The models of Domgörgen & Mathis (1994) predict a ratio of about 0.06. Heiles (1994) also finds lower limits on He ionization for the inner Galaxy below the model prediction from radio recombination line (RRL) observations of He and H. Reynolds therefore does not rule out other ways of producing the diffuse, high-z emission, and the problem of matching the observed line ratios is just one of the interesting issues addressed in several recent theoretical works on the origin and ionization of the WIM, which we turn to next.

3. New Theoretical Developments on the Origin and Maintenance of WIMs

3.1. IONIZATION STRUCTURE

Mathis' (1986) models of photoionization of diffuse gas by a dilute radiation field (the low-U case) were able to explain the enhanced [S II]/Hα, comparable [N II]/Hα, and low [O III]/Hβ found in the Reynolds layer relative to H II regions. However, recent limits on [O I]λ6300, more stringent limits on [O III]λ5007, and newly available line ratios for the WIM of NGC 891 (Dettmar & Schulz 1992; see Table 1) have led to reexaminations of the problem by Domgörgen & Mathis (1994) and Sokolowski (1994).

The models of Domgörgen & Mathis (1994) assume an ambient medium of uniform density and specified chemical composition, and a specified stellar spectrum from Kurucz (1979) of some temperature. Two other parameters are important for determining the ionization structure. The first is q, which is related to the ionization parameter and is defined as $q = nf^2L_{50}$, where L_{50} is the ionizing luminosity in units of 10^{50} photons s^{-1}. From observed emission measures (EMs) and DMs, q can be estimated to be 3×10^{-4} photons cm^{-3} s^{-1}. The second parameter is $X_{\mathbf{edge}}$, the ionization fraction at the edge of the model. Low $X_{\mathbf{edge}}$ corresponds to the matter bounded case, where some radiation escapes the galaxy completely. High $X_{\mathbf{edge}}$ corresponds to partially ionized surfaces of neutral clouds. While models with q given by its observationally constrained value best reproduce the observed line ratios of [O I], [O III], [S II] and [N II], no single best value of $X_{\mathbf{edge}}$ can be found. But a composite model with 80% "$X_{\mathbf{edge}} = 0.1$ gas " and 20% "$X_{\mathbf{edge}} = 0.95$ gas", which really represents a continuous range of $X_{\mathbf{edge}}$, reproduces all these ratios. A consequence of these constraints on $X_{\mathbf{edge}}$ is that some radiation escapes the Galaxy, a conclusion also found by Sokolowski (1994; see below) and Dove & Shull (1994; see below). The above-mentioned over-production of He I in the WIM could be avoided if the effective temperature of the ionizing spectrum were 36,000 K (corresponding to an O8 star) rather than their best value of 38,000 K (an O7 star), but with so late an O star dominating the integrated spectrum, it is not possible to produce a sufficient ionizing flux, according to Reynolds & Tufte (1994). Alternatively, an additional source of spectral softening in the intervening gas needs to be invoked to eliminate He-ionizing photons. Obviously, the O-star scenario still contains some unsolved mysteries.

Sokolowski (1994) is more concerned with reproducing the rather high ratios of [S II]/Hα and [N II]/Hα observed in NGC 891 and their increase with z (Table 1). A number of effects can help in creating these high line ratios in low-U cases: *a)* a mean interstellar opacity term which hardens the stellar radiation as it propagates upward; *b)* standard depletions of otherwise important coolants Si and Fe (and C and O to a lesser extent) onto dust grains, which shifts the burden of cooling to N and S; *c)* matter bounded models, which feature lower [O I]/Hα and [O III]/Hα ratios; and *d)* an upper cutoff to the stellar mass function above 40 $M_{\odot}$. The last feature implies that there are currently many very massive stars ionizing the gas in NGC 891 where the line ratios have been measured (the brightest region of the WIM at 1' NNE of the nucleus), i.e. this part of the disk is a region of elevated star formation activity. Other tracers (see below) also indicate that this is the most active star forming region of NGC 891. In contrast, the Reynolds layer requires an upper IMF cutoff somewhere below 40 $M_{\odot}$. The direct effect of dust on the thermal balance, through electron ejection and capture, is found to be small, in contrast to the conclusions of Reynolds & Cox (1992). However, if depletions are required, then there may be significant amounts of dust at heights $\sim$ 1 kpc from the midplane of NGC 891. Although vertical dust filaments can be seen (Keppel et al. 1991), it is not clear whether dust reaches sufficient heights. The observed rise of [S II]/Hα and [N II]/Hα with z is expected as a consequence of an ionization parameter falling rapidly with z in his models, which is a reasonable expectation if the filling factor does not change rapidly with z. For the WIM structure, the best models consist of matter bounded clouds of column density roughly 10^{19} cm^{-2}, and filling factors of about 0.1.

3.2. STRUCTURAL ISSUES, THE POROSITY PROBLEM, AND CHIMNEYS

two papers (Miller & Cox 1993, hereafter MC; Dove & Shull 1994, hereafter DS) have recently tackled the problem of whether ionizing photons from disk O stars can propagate far enough from their sources to explain the general properties of the Reynolds layer. DS are more concerned with simply calculating the fraction of Lyc photons which can escape the low-z gas and are available for ionization of the high-z gas. MC wish to understand the complete three-dimensional structure of the local WIM.

Both choose to defer inclusion of the complicating effects of holes and chimneys created by multiple stellar winds and supernovae, and solve a more tractable problem in which a simple, more uniform ambient ISM is considered. DS consider two models of the vertical structure of the ambient ISM: first, a single Gaussian component in hydrostatic equilibrium in the gravitational potential; second, the three-Gaussian distribution of Dickey & Lockman (1990); third, the second model plus a neutral component representing the Reynolds layer. MC consider a population of standard opaque clouds, specified by their mean free path and scale height, embedded in an extracloud medium, consisting of exponential components representing the Reynolds and Lockman layers.

For the ionizing photon flux, DS calculate the expected ionizing flux from a standard OB association, and then integrate this over an adopted luminosity function of such associations, normalized by the total Lyc flux per cm^2 of disk emitted by all known O-type stars in the Solar Neighborhood from a survey by Garmany, Conti, & Chiosi (1982). This survey is thought to be complete out to a distance of 2.5 kpc. DS assume that all of the photons are emitted at $z = 0$ pc, and distribute OB associations at random over a specified area of disk according to the luminosity function. MC actually adopt the spectral types and three-dimensional positions of all of the O stars in the Garmany *et al.* survey. Implicit in both of these approaches is the assumption that since these O stars are visible to us, they are not deeply embedded in molecular clouds and their Lyc is available to ionize large volumes of gas. Hence, we have a reasonable way of isolating the very O stars which are most likely responsible for the WIM ionization.

A key intermediate result from both groups is that a single O star sitting in a diffuse medium can ionize a volume hundreds of pc on a side. The volume can be elongated in z or even cone-shaped depending on the level of Lyc flux and the exact prescription of the vertically stratified ambient medium. Both groups also show face-on views of the resulting ionization structure due to the entire ensemble of O stars at various heights above the plane (Figure 1). At heights of several hundred pc, only a few ionization cones from the strongest Lyc sources exist, but the combined covering factor of the cones increases with z, at least in the model of DS. The models suggest that the WIM should be most prominent above active star forming regions.

DS find in their model with a four-component ambient ISM that an ionizing flux of 8×10^5 photons cm^{-2} s^{-1} escapes each side of the H I layer and penetrates the Reynolds layer – only 40% of the required flux. However, they do not consider that ionizations in the Dickey-Lockman layers could contribute to the WIM, although they admit this possibility. Hence, their models may reproduce roughly the necessary ionization requirement of the WIM. As an aside, they find that a flux of about 10^6 photons cm^{-2} s^{-1} escapes each side of the Galactic disk entirely. MC simply consider that the combination of all their extended ionization volumes *forms* the Reynolds layer. They are able to reproduce many of the integrated properties of the WIM, such as the mean midplane density, filling factor and the average value of $DM \times \sin|b|$ for pulsars and $EM \times \sin|b|$ for the $H\alpha$ background, all to within a factor of 2, as well as the observed deficit of ionized gas on the $+z$ side of the disk.

These models are, of course, very simplified, and it would be desirable to include effects such as possible incompleteness of the Garmany *et al.*

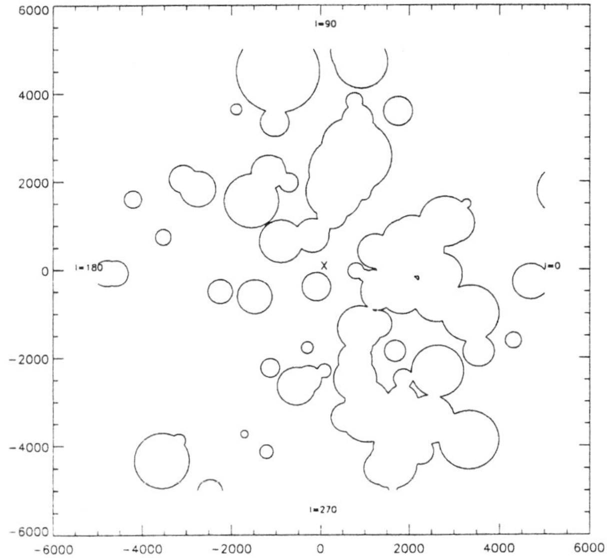

Figure 1. Top view of a cross-section at $z = +400$ pc through the standard model of Miller & Cox (1993). The Sun is at the center. Distances in the plane in pc and longitudes are indicated. The enclosed areas indicate the ionized regions at this height above the plane.

catalog, high density environments (~ 10 cm^{-3}) for some of the visible O stars, blister H II regions, a correct treatment of overlapping ionized regions, and other sources of ionizing photons. However, the rough agreement with the predicted and observed structure of the WIM is encouraging for the O-star photoionization picture. Most importantly, although the authors admit that chimneys are likely to be very important for the WIM and should be included in future models, their work implies that there is no real porosity problem because the visible O stars can keep the gas ionized.

Norman (1991) and Heiles (1992) stress the importance of worms and chimneys in understanding the WIM. The chimney model of the ISM was described by Norman & Ikeuchi (1989). Since chimneys open up large, evacuated regions in the ISM, they are natural candidates for creating absorption free paths in which ionizing photons can propagate upwards. Heiles' H I worms may well trace the walls of chimneys. Motivated by the aforementioned RRL observations, Heiles suggests that worms are closely related to the WIM, which he says could well stand for the "Worm Ionized Medium". Of the 118 Galactic worms catalogued by Koo, Heiles, & Reach (1992), which show vertical structure in H I , IRAS and 408 MHz maps, 56 are associated with large H II regions in the plane, and the authors claim that most of the 408 MHz emission from worms is thermal. Furthermore, two worms associated with H II regions have been detected in H166α. The line emission decreases rapidly with $|b|$. One of them, GW 31.6–5.9, has an ionization requirement 20 times larger than that of the H II region near its base, W43. If the cluster is doing the ionizing, then most of the radiation escapes its immediate environment and traverses large distances, as expected if many winds and supernovae have already cleared the path. Fourteen more worms have now been detected in RRL emission (Heiles 1994). Given these associations, Heiles proposes that the diffuse H166α emission surveyed by Lockman (1980) is associated with worms, shells and chimneys, and is the inner Galaxy version of the Reynolds layer. It remains to be seen how these two components compare in detail in properties such as filling factor and scale height.

Norman (1991) similarly proposes that superbubbles and chimneys feed the Reynolds and Lockman layers. The chimney walls are directly ionized by disk O stars below and should show a relatively hard spectrum similar to an H II region, and diffuse reradiation from the walls may ionize a much larger ($\sim$ 2 kpc diameter for a typical chimney) region, which would show a softer spectrum. It is important to remember that the OB stars that created the chimney are not the source of ionization. Either the timescale for the star formation event must be sufficiently long such that new stars are still forming after the chimney has been created, or else a second generation of star formation is required, perhaps triggered by the expanding bubble. The observed association of H II regions with H I worms suggests that a source of ionizing photons does exist in many cases, if worms trace chimneys. Norman, M. Voit and other collaborators are currently working on the details of radiative transfer in a chimney wall and the resulting spectrum of the ionized gas in the chimney and the surrounding diffuse gas (M. Voit, private communication).

3.3. OTHER THEORIES

The galactic fountain model, in which hot gas is raised into the lower halo by supernovae, cools and returns to the disk, has been around for many years (Shapiro & Field 1976; Bregman 1980). The chimney model can be considered as a refinement to the fountain model since it attempts to take into account the clumpy distribution of star formation in the disk. Shapiro & Benjamin (1993) are applying the fountain model to observations of halo UV emission lines of highly ionized species but are also testing the idea that the diffuse Hα emission could be due to cooling fountain gas. They believe that about 40% of the integrated emission column perpendicular to the disk could be explained in this way.

Slavin, Shull, & Begelman (1994) study "turbulent mixing layers", which are expected to occur at interfaces between hot and cold gas in the walls of superbubbles and chimneys. Shear flows at these boundaries cause hot and cold gas to mix, producing gas at temperatures of $10^{5.0-5.5}$K. This is another possible way of producing UV emission lines of highly ionized species, but also can give rise to optical line emission with ratios of [N II]/Hα (0.3–0.7) and [S II]/Hα (~ 0.5) about equal to those found in the Reynolds layer, NGC 891, and the froth in irregular galaxies. It is not clear whether all of the diffuse emission can be explained this way, but the authors believe it could account for >30% of the high-latitude emission from the Reynolds layer. Both this theory and the fountain theory tap the energy output by supernovae to explain the Reynolds layer.

More exotic is the idea that microflares from reconnecting magnetic field lines are a source of significant halo heating and ionization (Raymond 1992), and the theory that decaying dark matter could provide an important source of ionizing photons (Sciama 1990a,b,c). In this latter theory, one expects the production of $\sim$ 14 eV photons from the decay of ν_τ neutrinos. Dettmar & Schulz (1992) have shown, however, that such a monochromatic source of photons is unable to reproduce simultaneously all of the emission line ratios observed in NGC 891. Raising the photon energy to 18 eV helps, but then [O I] is over-produced and photons of this energy are not allowed by the theory anyway. Adding dust or lowering the metallicity does not seem to help the problem. Hence, as it currently stands, the theory cannot explain the bulk of the diffuse emission.

4. Observations of Diffuse Ionized Gas in External Galaxies

Observations of the Reynolds layer emission are hampered by absorption and our location within the layer. External galaxies can provide a more synoptic view of diffuse media, and hence in recent years much attention has turned to narrow-band imaging of edge-on spirals. Recent work on less highly inclined galaxies is also leading to some interesting results.

4.1. NGC 891

The first edge-on galaxy to be looked at was NGC 891, a nearby ($D = 9.5$ Mpc) edge-on ($i \gtrsim 88.6°$; Rupen 1991) with an optical appearance similar to the Milky Way. A prominent diffuse Hα halo (Figure 2) was revealed in observations by Rand, Kulkarni, & Hester (1990, hereafter RKH) and Dettmar (1990). A more recent image of one part of the disk is presented by Pildis, Bregman, & Schombert (1994a). Emission is detectable up to about 3 kpc above the plane. Modeling by RKH showed that at $R = 8.5$ kpc, the mean midplane density is roughly 0.05 cm^{-3}, the surface density is about 4 $M_{\odot}$pc^{-2}, the scale height is roughly 1 kpc (although evidence for a high-z tail to the emission was found), and the total mass of ionized gas is a few times 10^8 $M_{\odot}$. These quantities assume a filling factor of $\phi = 0.2$, and depend on $\phi^{1/2}$. In general, then, the WIM of NGC 891 is comparable to the Reynolds layer, with somewhat higher density and surface density at comparable radii. Comparing with H I data of Swaters (1994; see also Swaters *et al.* 1995, 1997; van der Hulst 1996), $\Sigma_{WIM}/\Sigma_{HI} = 0.8$ at this radius, whereas this ratio is equal to about 0.3 in the Solar Neighborhood (Kulkarni & Heiles 1988). Hence it seems a greater fraction of the ISM in NGC 891 is ionized. A higher star formation rate could result in more of the gas being ionized, and this is one of several indications for enhanced star formation (see below).

A very exciting result was the discovery of several worms of Hα emission, extending $\gtrsim$ 2 kpc off the plane, by both RKH and Dettmar (1990). Rand *et al.* interpreted these as ionized surfaces of chimney walls and concluded that the ISM of NGC 891 was an example of the chimney model (Norman & Ikeuchi 1989).

In light of the recent ionization calculations of DS and MC, it seemed worthwhile to apply a median filtering technique to the Hα image of RKH to remove the large-scale vertical gradient of the diffuse emission and see how the worms connect onto the plane. The median-filtered image was formed using a box of size 40$''$ along the major axis and 2$''$ along the minor axis, so that the local median z-gradient could be removed from every location without smearing out the gradient by averaging over a large region perpendicular to the plane. Dividing the original image by this median filtered image produces a "contrast image" which is shown in Figure 3 (one could also subtract the images, but the quotient brought out the filamentary structure better). The image has been rotated so that the major axis runs vertically, and blown up to show only the inner disk ($R < 2.75'$), where the diffuse emission is brightest. The worms stand out clearly. The two prominent worms in the upper left quadrant, for instance, can be traced from more than 2 kpc above the plane down to the midplane. Note that most of the worms have footprints in H II regions (seen as white) in the plane. This relationship is made more significant by the fact that there is substantial extinction in the plane (Keppel *et al.* 1991) which would tend to destroy such a relation by making many bright H II regions undetectable. Hence, the image suggests that the OB associations which power the H II regions also produce ionized vertical structures above them, i.e. the H II regions are very leaky and the photons can travel far. Whether the Hα worms are simply the ionization cones of DS or MC, or the chimney walls of Norman & Ikeuchi, is unclear, although morphologically some certainly have the very thin and tilted appearance one might expect for chimney

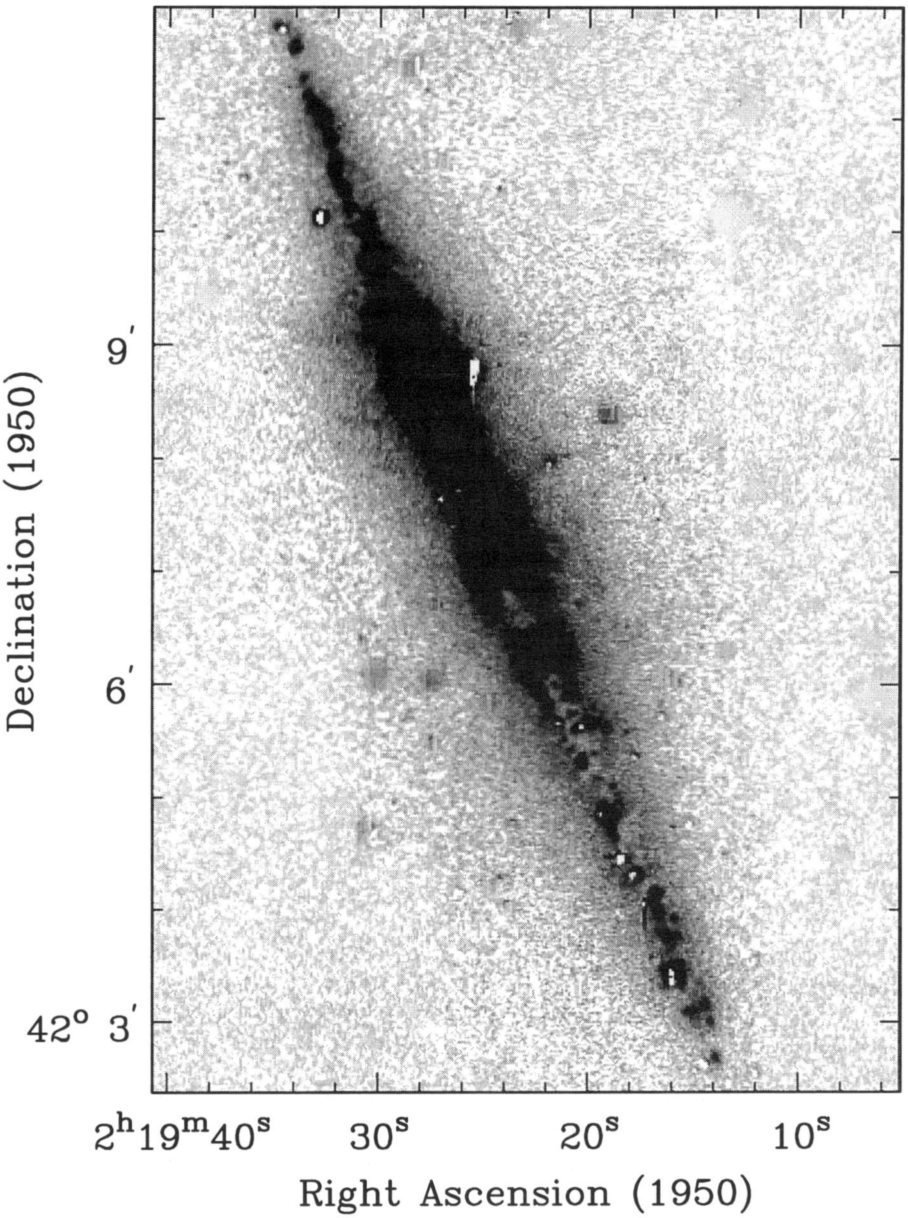

Figure 2. Diffuse Hα emission in NGC 891 from Rand, Kulkarni, & Hester (1990). The image has been stretched to show the faint emission.

walls, especially on the north side. One or two of the southern worms, though, have a shorter, broader appearance. Correlation or anticorrelation with H I structures could distinguish between these possibilities, but so far, comparisons have been inconclusive (R. Swaters, private communication). It should be pointed out that only several percent of the halo emission is contained in these worms. The bulk is in the more diffuse component removed from Figure 3.

A second tentative result reported here for the first time comes from comparing the [S II] and Hα images of RKH. The [S II]/Hα ratio is about a factor of 2 lower in the worms than in the surrounding gas. This would be consistent with the worms being illuminated with a harder, more HII-region-like spectrum, such that most of the S is doubly ionized, as expected in the picture of Norman (1991). However, this result relies on dividing one continuum-subtracted line image by another, and hence is rather sensitive to errors in the continuum subtraction, and needs to be confirmed.

Dettmar & Schulz (1992) carried out long-slit spectroscopy of the diffuse gas with the slit oriented perpendicular to the plane and centered at a location along the major axis above which the diffuse emission is very bright (at $R = 1'$ NNE of center). Ratios of [S II]/Hα and [N II]/Hα were found in the diffuse gas which are higher than in the Reynolds layer and increase with z (Table 1). RKH also found from [S II] and Hα images that the [S II]/Hα ratio increases with z up to about 500 pc, where it flattens out. As mentioned above, Sokolowski (1994) needed to invoke recent star formation to explain these high ratios, suggesting that at least this region is undergoing enhanced star formation. Spectra in a second region with fainter WIM emission, at $R = 40''$ SSW of the nucleus, show lower line ratios ([N II]/H$\alpha = 0.4 - 0.8$ and [S II]/H$\alpha = 0.25 - 0.35$ over the same z range). The increase of the ratios with z is expected in photoionization models as U decreases with height above the plane (i.e. distance from the O stars).

The WIM of NGC 891 is concentrated within $R = 8$ kpc (2.7'), and, particularly on the north side, shows a rather sharp cutoff at this radius. The emission is generally brighter on the north side and the brightest, most z-extended region occurs interior to the cutoff at $R = 1 - 2'$ (Figure 4), where some of the most prominent worms are found.

This galaxy also has a very bright and z-extended synchrotron halo (Dahlem, Dettmar, & Hummel 1994). The radial structure of the synchrotron halo emission shows the same characteristics as that of the Hα halo (Figure 4), suggesting that the two halos are intimately linked. Polarized radio emission at 6 cm in the halo is also brightest where the Hα and total radio halo emissions are brightest (Sukumar & Allen 1991). The polarization vectors at 6 cm, where Faraday depolarization effects are rela-

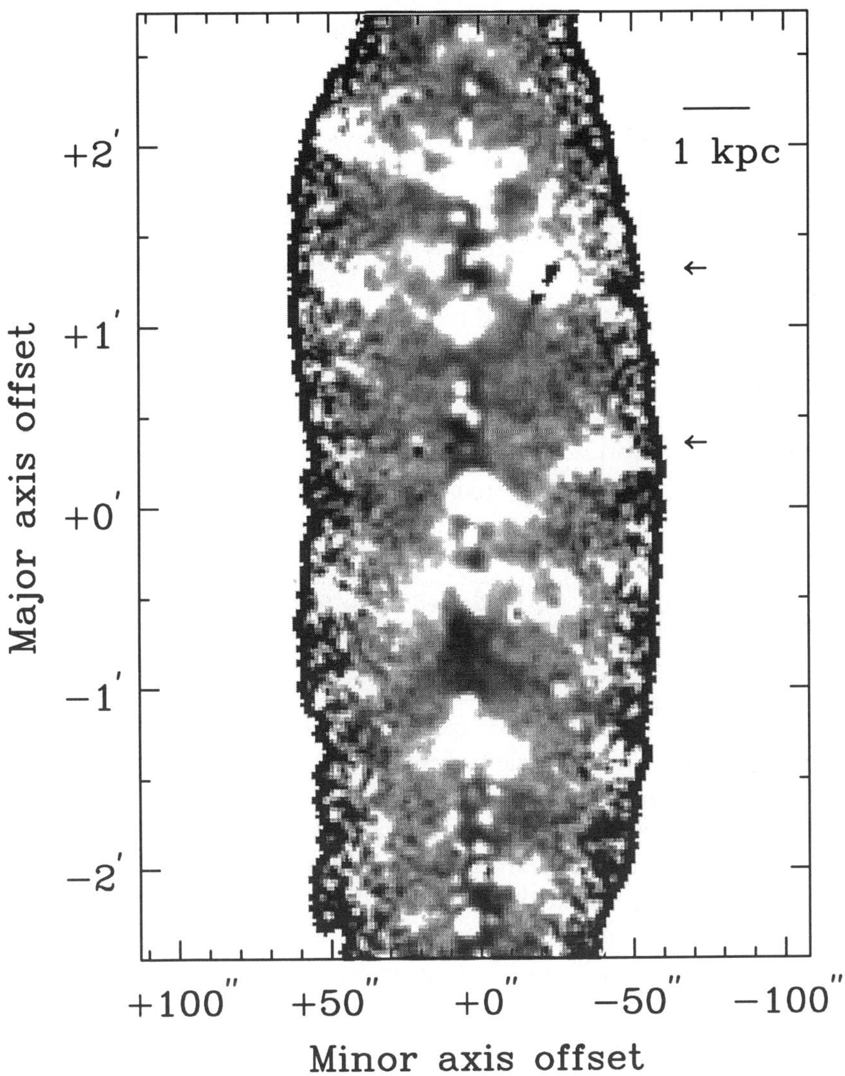

Figure 3. Image of the diffuse Hα emission in NGC 891 divided by a median-filtered image to bring out the filamentary structure. Only the inner disk is shown, where the halo emission is brightest. Filaments show up in white against a grey background. Note that most filaments connect onto HII regions in the disk. Also note the complete loop on the south-west side of the disk, centered about 1 kpc from the midplane The bright object of about 20″ size with the dark band across it indicated by the upper arrow is a foreground star. Another foreground star is blended with the filament indicated by the lower arrow. The dark edge of the image is a processing artifact.

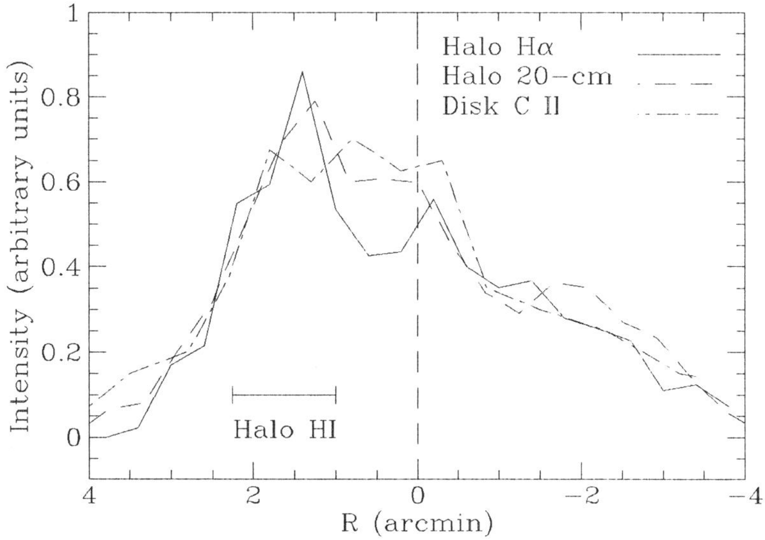

Figure 4. Radial profiles of Hα halo (at 30″ above the plane on the east side), synchrotron halo (at 45″ above the plane on the east side), and the disk [C II] emission in NGC 891. The prominent region of high-z HI discussed in the text is indicated by the bar. Positive offsets refer to the north side. Note the similarity of the three profiles.

tively small, indicate a well-ordered, vertically oriented field, suggesting that an outflow from the disk is driving the field into this configuration. Further evidence for a wind comes from Hummel *et al.* (1991), who conclude that the complicated structure of the halo as well as significant flattening of the spectral index with z are best explained by convection of the cosmic rays by a wind from the disk. A purely diffusive halo would have a much smoother and more uniform shape.

Strong evidence that star formation is responsible for the production of the halos comes from a comparison of these profiles with the midplane major axis profile of [C II] 158μm emission (Figure 4). This emission arises from photon-dominated regions (PDRs) of molecular clouds in star-forming regions, and hence provides a more direct tracer of ongoing star formation in the disk (Madden *et al.* 1993, 1995). The radial profile of [C II] emission shows the same asymmetries as the two halo profiles. The higher line ratios in the WIM are also associated with the region of the disk which is more active in forming stars.

Has star formation simply resulted in more ionization of high-z gas, or is there any evidence that star formation has actually raised gas off the plane? Recent H I data suggests that there is. Swaters (1994), using very sensitive 21-cm data from the Westerbork Synthesis Radio Telescope, has analyzed the velocity structure of high-z gas in NGC 891 to determine whether there is a component which can be associated with the inner disk and separated from outer disk phenomena such as flares and warps. Using a mask in position-velocity diagrams parallel to the major axis, he can reject outer disk gas along lines of sight which pass close to the center of the galaxy, and produce a map of purely inner disk gas (Figure 5). Emission close to the systemic velocity near the center of the galaxy is not included in the mask (as it could come from anywhere along the line of sight), hence the apparent central hole in Figure 5. Swaters indeed finds such a high-z, inner disk component, and it is concentrated at the position where the Hα and radio halos and [C II] disk emission are brightest, i.e. at $R = 1 - 2'$ NNE of the nucleus, as indicated in Figure 4. H I emission here is detected up to 2.5 kpc from the midplane.

An attractive interpretation ties all of these correlations together. Strong star formation has lifted gas off the plane by creating superbubbles and chimneys. Some of the chimneys show up as Hα worms, ionized by O stars below. The chimneys allow efficient upward propagation of cosmic rays, leading to the synchrotron halo. All of these processes are most active at $R = 1 - 2'$ on the NNE side.

The mass of the WIM is roughly a few times 10^8 $M_\odot$. In the chimney model of Norman & Ikeuchi, a mass flow rate through the halo of 0.3–3 $M_\odot$ yr^{-1} is expected in the Milky Way case, and the dynamical timescale is a few times 10^7 yr. Even if all of this mass flow were in the form of WIM, the predicted mass in the halo would fall short by an order of magnitude. Much more vigorous chimney action would be required in NGC 891. However, the derived WIM mass is model dependent and varies as the square root of the filling factor, and could well be lower.

More observational constraints on the flow of hot gas come from the discovery of a diffuse X-ray halo with ROSAT (Bregman & Pildis 1992). The scale height of the hot gas is about 7 kpc, much greater than that of the warm ionized component. There is weak evidence that the gas is bound to the galaxy. In a fountain model, the emission implies a modest cooling rate of 0.12 $M_\odot$ yr^{-1}. Ballistically falling cooled X-ray gas can therefore only account for a small fraction of the warm gas mass – 3×10^6 $M_\odot$. This is an argument in favor of ionization of the WIM by O stars, unless the X-ray gas can be prevented from falling ballistically. The radial extent of the emission is almost as large as that of the bright Hα and radio halo components, but does not seem to show much evidence for a north-south asymmetry.

What is missing from the picture is kinematic evidence for an outflow. However, this is difficult to infer in edge-on galaxies since the vertical

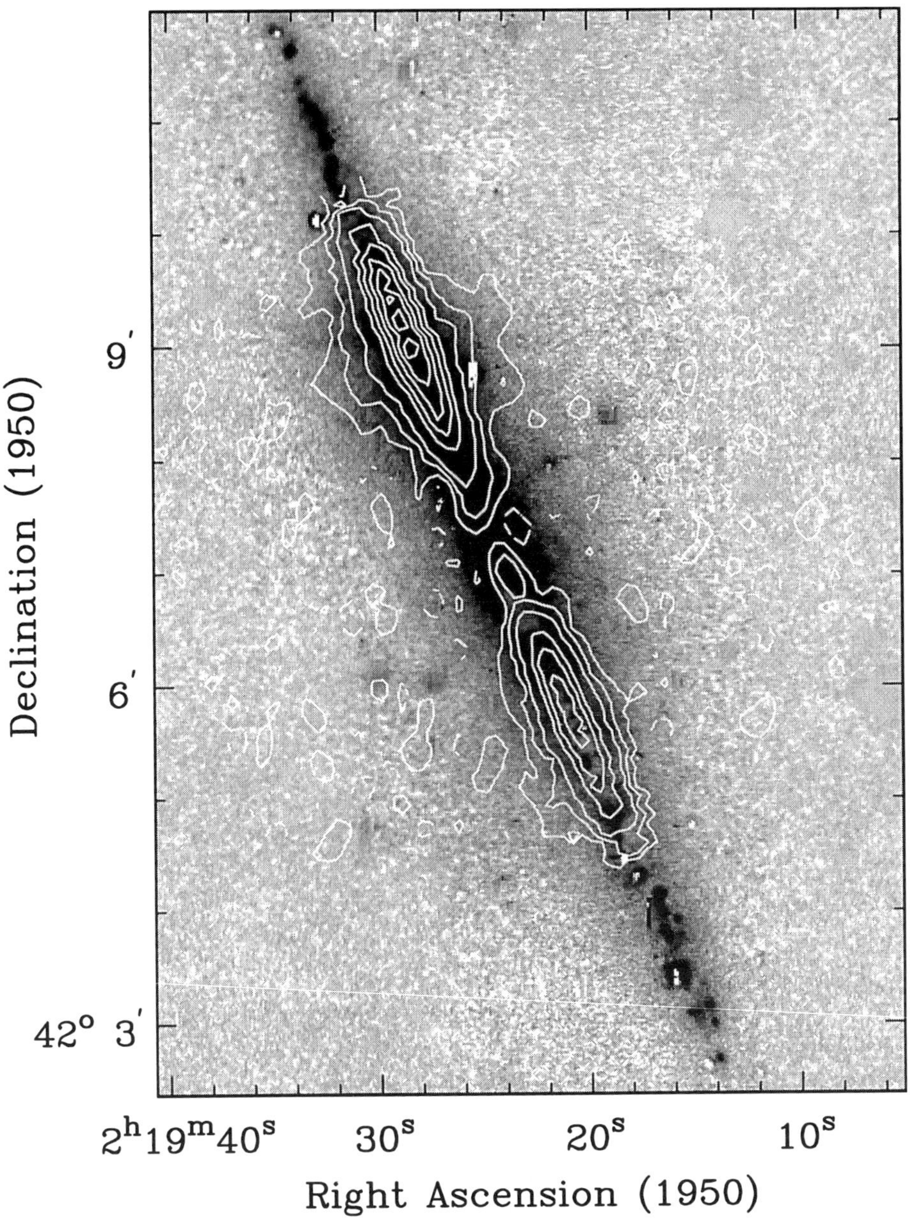

Figure 5. Map of HI in the inner disk of NGC 891 from Swaters (1994) overlaid on the Hα image. Note that the prominent region of high-z HI coincides in radius with the most prominent part of the Hα halo.

motion is transverse to the line of sight, and the large velocity width along any line of sight through the disk due to galactic rotation makes detection of any in-plane peculiar motions unlikely. However, we do have ample kinematic evidence in face-on spirals for high-velocity gas associated with holes in the disk (e.g. Kamphuis 1994), with the large, two-sided expanding hole in M101 reported by Kamphuis, Sancisi, & van der Hulst (1991) presenting an excellent case for a supernovae-driven outflow. Schulman, Bregman, & Roberts (1994) find evidence for high-velocity wings, which are unlikely to be due to warps, in integrated H I profiles of many galaxies in an optically selected sample.

Finally, although NGC 891 is an excellent site to study galactic WIMs, it should be pointed out that the star formation rate may be a few times higher than in the Milky Way, as indicated by the global luminosity of tracers which are at least in part related to ongoing star formation: CO emission (Scoville *et al.* 1992), 20-cm emission (Allen, Baldwin, & Sancisi 1978), and FIR emission (Wainscoat, de Jong, & Wesselius 1987). Hence, a generally enhanced star formation rate may be the reason for the prominent z-extended layer of diffuse ionized gas. These quantities would all be more modest, of course, if we are using a distance too large by, say, a factor of 2. However, a recent analysis based on planetary nebula luminosity functions suggests that the distance is 9.9 ± 0.7 Mpc (Ciardullo *et al.* 1991), close to the adopted distance.

4.2. OTHER EDGE-ON GALAXIES

Several other edge-on galaxies have now been observed in Hα with a variety of results.

NGC 4631 (Rand, Kulkarni, & Hester 1992) is a disturbed, interacting galaxy which is bright in H II regions (Figure 6). Patchy, diffuse halo Hα emission was found above the central few kpc of the major axis. Also found were two well-defined worms above the nucleus, extending more than 2 kpc from the plane and separated by about 1 kpc along the major axis, which together suggest a conical structure and probably represent a blown out superbubble due to a nuclear starburst. Morphologically, the worms suggest the Hα loop above the nucleus of NGC 3079 (Ford *et al.* 1985) but seen at a later stage of the evolution of the superbubble. Also very noticeable in Figure 6 is a major disturbance to the star forming layer towards the east end of the disk, where many H II regions are found well off the plane. A very large H I supershell is coincident with this disturbance (Rand & van der Hulst 1993), and apparently star formation has occurred in the supershell at high-z. Since our interest here is in extraplanar *diffuse* gas, we do not discuss this feature further here. Wang *et al.* (1994) also

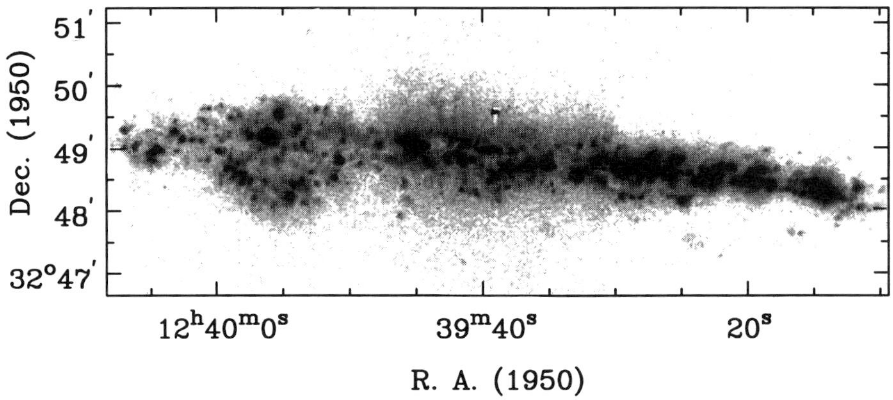

Figure 6. Diffuse Hα emission in NGC 4631 from Rand, Kulkarni, & Hester (1992). A logarithmic stretch has been used to show both the diffuse gas and the HII regions.

present an Hα image of NGC 4631.

This galaxy has X-ray and radio halos even brighter and more extended than in NGC 891 (Wang *et al.* 1994; Hummel & Dettmar 1990; Trinchiery, chapter 5). Both halos show a concentration to the innermost few kpc of the major axis and are brighter on the north side, where the pair of Hα worms were found. Again, the radio halo structure is best understood by convection due to a wind from the disk. The magnetic field in the halo is well studied, and the derived vectors indicate a conical field morphology which points back to the center of the galaxy. Such a configuration is expected for a field shaped by a wind from the disk (Brandenburg *et al.* 1992). As pointed out by Wang *et al.* (1994), the halos in NGC 4631 could be more extended than those in NGC 891 because the outflow is occurring in a lower gravitational potential (escape speeds are about 340 km s^{-1} and 510 km s^{-1}, respectively). Some of the X-ray emitting gas may escape NGC 4631 altogether, according to Wang *et al.* A kinematic detection of a wind from NGC 4631 would be very interesting. Donahue *et al.* (1995) find evidence for some very faint patches of Hα emission some 16 kpc from the plane of NGC 4631, possibly associated with the tidal features between NGC 4631 and NGC 4656.

More evidence for a central starburst comes from the bright CO emission within about 3 kpc of the nucleus (Golla & Wielebinski 1994), which suggests a pile-up of molecular gas, possibly driven by the encounters the galaxy is undergoing, as in models by, e.g., Noguchi (1988).

A more extensive Hα halo is seen in the edge-on starburst galaxy NGC 5775 (Dettmar 1992). Unlike NGC 891, the emission is more confined to filamentary structures, and these filaments can extend to 6 kpc from the midplane. But like NGC 891, the halo emission is seen in all four quadrants and over a large range of the major axis. The Hα filaments show a correlation with spurs in a 20-cm map. NGC 2188 is a southern late-type edge-on spiral which has been observed in Hα by R.-J. Dettmar (private communication). A region of intense star formation in the disk, obvious even in broad-band images, turns out to be the most prominent region of diffuse Hα emission, featuring many vertical filaments which join onto disk H II regions, as in NGC 891.

Other edge-ons do not show anything like these three. NGC 4565 (Rand *et al.* 1992), NGC 4244 (Walterbos 1991), NGC 4302, NGC 5023 and NGC 2820 (work in progress by R. Rand and J. M. van der Hulst), and six galaxies observed by Pildis, Bregman, & Schombert (1994b) show either no detectable halo emission, or just one or two plumes. This picture is confirmed by a more recent study by Rand (1996) who presents Hα imaging of nine more galaxies (UGC 4278, NGC 4013, NGC 4217, NGC 4302, NGC 4762, NGC 5023, NGC 5746, NGC 5907 and UGC 10288. Only NGC 4302 shows vertically extended Hα emission over a large fraction of the disk (the central 8 kpc) extending up to $z \sim 2$ kpc.

4.3. A TREND LINKING THE WIM TO STAR FORMATION

The governing factor for extended Hα halos seems to be star formation activity, as first pointed out by Dettmar (1992) and Rand *et al.* (1992). Table 2 is adapted and extended from a similar table by Dettmar (1992) and shows that only the galaxies with L_{FIR} equal to the Milky Way value or greater show evidence for extensive Hα halo emission (L_{FIR} is probably not a perfect tracer of the star formation rate [e.g. Sauvage & Thuan 1992; Sauvage, chapter 1] but Hα cannot be relied upon in edge-ons because of extinction). The result is similar if the galaxies are ranked by total radio power, although this could be expected from the well-known FIR-radio correlation. Not mentioned in Table 2 are the facts that the three external galaxies with extensive Hα halo emission are also the ones with bright, extended radio halos, and that NGC 891 and NGC 4631, at least, have extended diffuse X-ray halos. However, many of the galaxies have not been observed in radio continuum or X-rays.

4.4. FACE-ON GALAXIES

M31 is, of course, not a face-on galaxy, but it is so close that is is possible to study the two-dimensional distribution of diffuse emission over the disk. Recent Hα and [S II] imaging of this galaxy by Walterbos & Braun (1994) have revealed a WIM which is associated with the spiral arms and concentra-

TABLE 2. WIM and Star Formation in Edge-on Galaxies

Galaxy	D (Mpc)	L_{FIR} (10^{43} erg s^{-1})	DIG morphol.	DIG: h_e (kpc)	1.4 GHz power (10^{21} W/Hz)
NGC 5775	26.7	9.2	many filaments	~ 1	18
NGC 891	9.5	5.0	diffuse+filaments	0.6–1.0	7
Milky Way	-	3.8	diffuse+filaments	0.9	3
NGC 4631	7.5	3.6	diffuse,patchy	0.2–1	8
NGC 4302	16.8	$\lesssim$ 1.2	2 diffuse plumes	-	?
NGC 4565	9.7	1.0	no halo DIG	-	1
NGC 2820	18	0.9	2 plumes	-	2.5
Six galaxies of Pildis *et al.* 1994b	25–72	0.4–1.6	0, 1 or 2 plumes	-	?
NGC 4244	3.1	0.04	no halo DIG	-	<0.05
NGC 5023	6	<0.01	no halo DIG	-	?

ted around star forming regions, extending a few hundred parsecs from the H II regions. The medium is characterized by filaments, shells, and diffuse gas. The diffuse emission accounts for 40% of the total Hα luminosity of M31, uncorrected for extinction. The average EM perpendicular to the disk ranges from about 6 to 15 pc cm^{-6}, whereas in the Solar Neighborhood this number is only 4.5 pc cm^{-6} (Reynolds 1989). However, since this average refers only to the small fraction of the disk which shows diffuse emission, only a modest star formation rate is required to maintain the layer.

Another important result for understanding the origin of the WIM is the behavior of the [S II]/Hα ratio as a function of distance from H II regions. Figure 7 shows a profile of [S II] and Hα emission for a cut across the spiral arm. The cut crosses a few H II regions, which show a low [S II]/Hα ratio. The ratio smoothly increases with distance away from the H II regions, reaching a typical value of 0.5 in the diffuse gas. This is exactly what is expected if the clusters producing the H II regions are also ionizing the diffuse gas – the radiation field becomes more dilute with distance from the cluster (U decreases), favoring the lower ionization state of sulfur. The general concentration of the WIM around H II regions also supports the O-star photoionization idea. A recent study of galaxies in the Sculptor Group confirms this picture (Hoopes *et al.* 1996).

M33 was the first external galaxy in which diffuse ionized gas was found (Carranza *et al.* 1968). This galaxy has been observed more recently by

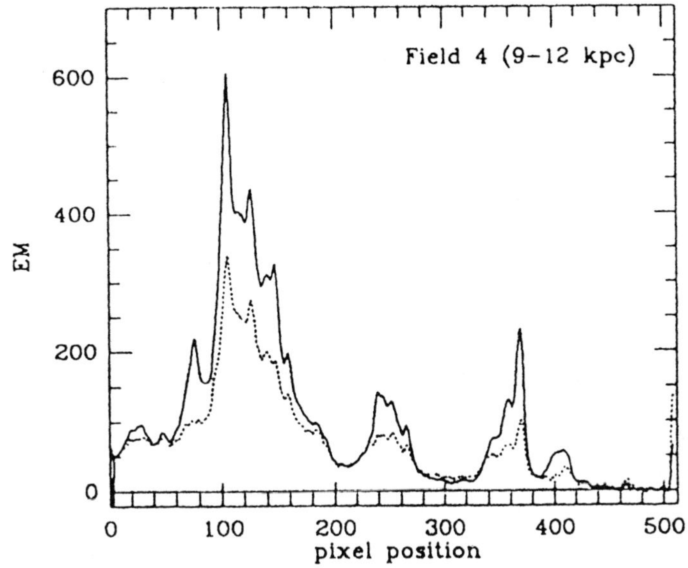

Figure 7. Profile of Hα (solid line) and [S II] (dotted line) emission across an arm of M31. The [S II] intensity has been multiplied by a factor of 2. Moving from bright H II regions into diffuse gas, the [S II]/Hα ratio smoothly increases to about 0.5.

Hester & Kulkarni (1990), and the morphology of the emission they see suggests filaments and sheets as well as diffuse gas.

Spectroscopy of eight shells of ionized gas in M33 and two in M31 by Hunter (1994) reveal H II region-like spectra, with [O III]/Hβ ~ 1 (but with a large spread), i.e. much larger than in diffuse gas, and [S II]/Hα = 0.1–0.4. These shells are typically 200 pc in diameter and have OB associations within them. This contrast in spectra between shells and truly diffuse gas calls to mind the predictions of Norman (1991) for ionized chimney walls discussed above.

It will be very interesting to compare H I holes with diffuse ionized gas features. These have been extensively catalogued in M31 by Brinks & Bajaja (1986) and in M33 by Deul & den Hartog (1990), and number about 150 in each galaxy. Are H I holes also devoid of diffuse ionized gas, as expected if the regions have been swept clean by winds and supernovae, or are some of the H I holes simply due to near-complete ionization by an OB association, as in the MC and DS models, and filled with diffuse gas? Is there a difference in this respect between H I holes that show expansion and those that don't, or the sizes of the holes?

HII regions cannot explain the entire Hα luminosity of M51, and it is likely that a significant fraction of the Hα emission is from a diffuse medium

(Rand 1992). R. Walterbos (private communication) also finds evidence for diffuse gas in M51.

4.5. IS THE WIM A GENERAL FEATURE OF GALAXY ISMS?

Most of the edge-ons observed so far do not show extended, diffuse *extraplanar* emission, yet M31, a galaxy with an L_{FIR} comparable to these null-result edge-ons, shows a WIM in its disk. Hence it seems premature to conclude from the edge-ons that only the most active star-formers (at the Milky Way level or higher) have WIMs. It may be that these few galaxies are the only ones with sufficient activity to push large amounts of gas to heights greater than ~ 1 kpc.

Evidence that spiral galaxies in general have WIMs comes from their integrated spectra (Lehnert & Heckman 1994). These authors have looked at line ratios such as [N II]/Hα and [S II]/Hα for a subset of galaxies studied by Kennicutt (1992) with Hα equivalent widths > 30Å . This restriction is designed to reject galaxies which might have line intensities affected by underlying stellar absorption, although it also creates a heavy bias toward late-type spirals and a slight bias against luminous galaxies. They find that the integrated [S II]/Hα ratios are typically 20%–100% higher than seen in H II regions while [N II]/Hα averages about 25% higher, implying that a significant fraction of the Hα flux arises from diffuse gas in these galaxies. Hardly any of the galaxies have known AGNs (which could affect the global ratios) and the authors argue that unrecognized LINERs in the sample would not affect the integrated line ratios severely. It will be interesting to image some of these galaxies so that the global line ratios can be related to other properties of their WIMs. It is also important to understand whether the sample is biased to galaxies with high star formation rates.

5. Conclusions

Evidence is emerging that WIMs are a general feature of spiral galaxies, and that ionization by disk O-stars is the most viable way of maintaining them. Other mechanisms may contribute to the WIM emission, but are probably not the dominant ones. The role of chimneys, the exact ionization structure of the WIM, and the details of radiative transfer are not yet fully understood. Diffuse extraplanar gas seems to occur only in the most actively star-forming edge-ons, is usually concentrated to the inner disk, and is correlated with the presence (and in some cases the detailed structure of) radio halos, X-ray halos and high-z H I .

During the course of preparing this review, I have benefited from many useful discussions with M. Dahlem, R.-J. Dettmar, C. Heiles, J. M. van der Hulst, M. Lehnert, R. Reynolds, R. Sancisi, M. Voit and R. Walterbos. Helpful comments on the manuscript by S. Vogel are appreciated. I thank M. Dahlem for providing the 20-cm data for Figure 4.

References

Allen, R. J., Baldwin, J. E., and Sancisi, R. 1978, *Ast.Ap.* , 62, 397.

Brandenburg, A., Donner, K. J., Moss, D., Shukurov, A., Sokoloff, D. D., and Tuominen, I. 1993 *Ast.Ap.* 271, 36.

Bregman, J. N. 1980, *Ap.J.* , 236, 577.

Bregman, J. N. and Pildis, R. A. 1994, *Ap.J.* , 420, 570.

Brinks, E., and Bajaja, E. 1986, *Ast.Ap.* , 169, 14.

Carranza, G., Courtès, G., Georgelin, Y. P., Monnet, G., Pourcelot, A. 1968, *Ast.Ap.* , 31, 68.

Ciardullo, R., Jacoby, G. H., Harris, W. E. 1991, *Ap.J.* , 383, 487.

Dahlem, M., Dettmar, R.-J., and Hummel, E. 1994 *Ast.Ap.* , 290, 384.

Dettmar, R.-J. 1990, *Ast.Ap.* , 232, L15.

Dettmar, R.-J. 1992, *Fund. Cosmic Phys.* , 15, 143.

Dettmar, R.-J. and Schulz, H. 1992, *Ast.Ap.* , 254, L25.

Deul, E. R., and den Hartog, R. H. 1990, *Ast.Ap.* , 229, 362.

Dickey, J. M., and Lockman, F. J. 1990, *Ann.Rev.Ast.Ap.* , 28, 215.

Domgörgen, H., and Mathis, J. S. 1994, *Ap.J.* , 428, 647.

Donahue, M., Aldering, G. and Stocke, J.T. 1995 *Ap.J.* , 450, L45.

Dove, J. D., and Shull, J. M. 1994, *Ap.J.* , 430, 222 (DS).

Ford, H. C., Dahari, O., Jacoby, G. H., Crane, P. C., and Ciardullo, R. 1986, *Ap.J.* , 311, L7.

Garmany, C. D., Conti, P. S., and Chiosi, C. 1982, *Ap.J.* , 263, 777.

Golla, G. and Wielebinski, R. 1994 *Ast.Ap.* , 286, 733.

Heiles, C. 1992, in *Evolution of Interstellar Matter and Dynamics of Galaxies*, eds J. Palouŝ, W. B. Burton, and P. O. Lindblad, Cambridge Univ. Press, p. 12.

Heiles, C. 1994, in *Physics of the Gaseous and Stellar Disks of the Galaxy*, ed. I. R. King (ASP Conf Series 66), p 249.

Hester, J. J., and Kulkarni, S. R. 1990, in *The Interstellar Medium in External Galaxies: Summaries of Contributed Papers*, eds D. J. Hollenbach and H. A. Thronson, NASA Conf. Publ. 3084, p. 288.

Hoopes, C.G., Walterbos, R.A.M. and Greenwalt, B.E. 1996, *A.J.* , 112, 1429.

Hummel, E., Beck, R., and Dahlem, M. 1991, *Ast.Ap.* , 248, 23.

Hummel, E., and Dettmar, R.-J. 1990, *Ast.Ap.* , 236, 33.

Hunter, D. A. 1994, *A.J.* , 108, 1659.

Hunter, D. A., and Gallagher, J. S. 1990, *Ap.J.* , 362, 480.

Kamphuis, J. J. 1994, PhD Thesis, Univ. of Groningen.

Kamphuis, J. J., Sancisi, R., and van der Hulst, J. M. 1991, *Ast.Ap.* , 244, L29.

Kennicutt, R. C. 1992, *Ap.J.* , 388, 310.

Keppel, J. W., Dettmar, R.-J., Gallagher, J. S., Roberts, M. S. 1991, *Ap.J.* , 374, 507.

Koo, B.-C., Heiles, C., and Reach, W. T. 1992, *Ap.J.* , 390, 108.

Kulkarni, S. R., and Heiles, C. E. 1988, in Galactic and Extragalactic Radio Astronomy, ed. G. L. Verschuur and K. I. Kellermann (New York: Springer-Verlag), 95.

Kurucz, R. L. 1979, *Ap.J.Suppl.* , 40, 1.

Lehnert, M. D., and Heckman, T. M. 1994, *Ap.J.* , 426, L27.

Lockman, F. J. 1908, in *Radio Recombination Lines*, ed. P. A. Shaver (Dordrecht: Reidel), p. 185.

Madden, S. C. *et al.* 1993, in *The Evolution of Galaxies and Their Environment, The Contributed Papers*, eds. D. Hollenbach, J. M. Shull and H. A. Thronson, p. 185.

Madden, S. C., Geis, N., Townes, C. H., Genzel, R., Hermann, F., Poglitsch A. and Stacey 1995, in *Airborne Astronomy Symposium on the Galactic Ecosystem: From Gas to Stars to Dust*, ASP Conf. nr. 73, p. 181.
Mathis, J. S. 1986, *Ap.J.* , 301, 423.
Miller, W. W., and Cox, D. P. 1993, *Ap.J.* , 417, 579 (MC).
Noguchi, M. 1988, *Ast.Ap.* , 228, 635.
Norman, C. A., 1991 in *The Interstellar Disk-halo Connection in Galaxies*, IAU Symp. 144, ed. H. Bloemen, 337.
Norman, C. A., and Ikeuchi, S. 1989, *Ap.J.* , 345, 372.
Pildis, R. A., Bregman, J. N., and Schombert, J. M. 1994a, *Ap.J.* , 423, 190.
Pildis, R. A., Bregman, J. N., and Schombert, J. M. 1994b, *Ap.J.* , 427, 160.
Rand, R. J. 1992, *A.J.* , 103, 815.
Rand, R. J. 1996, *Ap.J.* , 462, 712.
Rand, R. J., Kulkarni, S. R., and Hester, J. J. 1990, *Ap.J.* , 352, L1 (RKH).
Rand, R. J., Kulkarni, S. R., and Hester, J. J. 1992, *Ap.J.* , 396, 97.
Rand, R. J., and van der Hulst, J. M. 1993, *A.J.* , 105, 2098.
Raymond, J. C. 1992, *Ap.J.* , 384, 502.
Reynolds, R. J. 1989, in *Galactic and Extragalactic Background Radiation*, IAU Symp. 139, ed. S. Bowyer and C. Leinert, 157.
Reynolds, R. J. 1993, in *Back to the Galaxy*, eds. S. S. Holt and F. Verter (A.I.P. Conf. Proc. 278), p. 156.
Reynolds, R. J., and Cox, D. P. 1992, *Ap.J.* , 400, L33.
Reynolds, R. J., Roesler, F. L., Scherb, F., and Boldt, E. 1971, in *The Gum Nebula and Related Problems*, eds. S. P. Maran, J. C. Brandt, and T. P. Stecher (NASA SP-332), p. 169.
Reynolds, R. J., and Tufte, S. L. 1995, *Ap.J.* , 448, 715.
Rupen, M. P. 1991, *A.J.* , 102, 48.
Sauvage, M. and Thuan, T. X. 1992, *Ap.J.* , 396, L69.
Schulman, E., Bregman, J. N., and Roberts, M. S. 1994, *Ap.J.* , 423, 180.
Sciama, D. W. 1990a, *Ap.J.* , 364, 549.
Sciama, D. W. 1990b, *Phys. Rev. Lett.*, 65, 2839.
Sciama, D. W. 1990c, *Nature* , 346, 40.
Scoville, N. Z., Thakkar, D., Carlstrom, J. E., and Sargent, A. I. 1993, *Ap.J.* , 404, L59.
Shapiro, P. R., and Benjamin, R. A. 1993, in *Star Formation, Galaxies, and the Interstellar Medium*, eds. J. Franco, F. Ferrini, and G. Tenorio-Tagle (Cambridge Univ. Press), p. 275.
Shapiro, P. R., and Field, G. B. 1976, *Ap.J.* , 205, 762.
Slavin, J. D. Shull, J. M. and Begelman, M. C. 1993, *Ap.J.* , 407, 83.
Sokolowski, J. 1994, preprint.
Sokolowski, J. and Bland-Hawthorn, J. 1993, in *The Evolution of Galaxies and Their Environment*, NASA Conf, Pub. 3190, ed. Hollenbach, D, Thronson, H, ans Shull, J.M., p. 324.
Sukumar, S., and Allen, R. J. 1991, *Ap.J.* , 382, 100.
Swaters, R. 1994, Undergraduate Thesis, Univ. of Groningen
Swaters, R., Sancisi, R. and van der Hulst, J.M. 1995 Astro Lett. and Communications, **31**, 161, 1995.
Swaters, R., Sancisi, R. and van der Hulst, J.M. 1997, *Ap.J.* , submitted
van der Hulst, J.M. 1996, in *Minnesota lecture series on H*I *in the Universe*, ed. E.D. Skillman, ASP. Conf. Series. p. 47.
Wainscoat, R. J., de Jong, T., Wesselius, P. R. 1987, *Ast.Ap.* , 181, 225.
Walterbos, R. A. M. 1991, in *The Interstellar Disk-halo Connection in Galaxies*, IAU Symp. 144, ed. H. Bloemen, 223.
Walterbos, R. A. M. and Braun, R. 1994, *Ap.J.* , 431, 156.
Wang, Q. D., Walterbos, R. A. M., Steakley, M. F., Norman, C. A., and Braun, R. 1995, *Ap.J.* , 439, 176.

HOT GAS AND X-RAYS

G. TRINCHIERI
Osservatorio Astronomico di Brera
via Brera 28
20121 Milano,
Italy

1. Is there Hot Gas in Normal Galaxies?

Detailed X-ray observations of normal galaxies started with the data from the imaging instruments aboard the *Einstein* Observatory. Several papers and review articles have been written on the basic results of these observations (see Fabbiano 1989; Fabbiano, Kim & Trinchieri 1992; Kim, Fabbiano & Trinchieri 1992a,b; among others), from which it is clear that different mechanisms for X-ray emission are active in normal galaxies and contribute in various degrees to the overall emission from these objects.

Major differences are found in the dominating mechanism producing the X-ray luminosities in different types of galaxies. Spiral and irregular galaxies contain a population of individual sources (mostly massive and low mass binaries and supernova remnants) whose collective emission is what is detected. The X-ray images of the sources associated with these objects clearly show peaks of emission in the bulge, disk, spiral arms and in the globular cluster population, with luminosities reaching $L_X \sim 10^{38-40}$ erg s^{-1} each. The variability of some of these supports their identification with single sources. Fainter individual sources, down to the coronal emission of single stars, are also present, but they cannot be individually detected and their collective contribution does not dominate in the overall balance. The spectral signature of the emission as a whole is also consistent with a population of individual, mostly hard sources in these galaxies.

Therefore the traditional view that late type galaxies contain large amounts of gas, and that this gas could be found in different phases, from the cold H I emission to the hot halos, was not supported by the results from the *Einstein* data. In fact, the X-ray emission could *possibly* be due to hot

J. M. van der Hulst (ed.), The Interstellar Medium in Galaxies, 129–151.

gas only in a few normal late type galaxies, and in all cases alternatives were also just as likely. Hot gas was suggested to explain the soft disk emission of NGC 4631 (Fabbiano & Trinchieri 1987; Vogler & Pietsch, 1996), and in M33 (Trinchieri *et al.* 1988). The data, however, would not allow a precise determination of the soft X-ray luminosity, leaving open the possibility that a population of individual sources, like low mass stars, could explain the presence of this component without the necessity of a hot gas component.

Hot gas was instead clearly detected associated with starburst activity. The extended, extraplanar emission observed in the X-ray images of M82, NGC 253 and NGC 3628 for example (Watson *et al.* 1984; Fabbiano & Trinchieri 1984; Fabbiano 1988; Fabbiano *et al.* 1990) has been interpreted as hot gas flowing out from the starburst nuclei of these galaxies.

In contrast to the late type galaxy population, hot gas has indeed been detected in the early types, traditionally thought to be devoid of gas. In particular, the hot gas component dominates the emission in the brightest objects, as suggested by the excess X-ray luminosity over the expected contribution from a binary population, their distorted X-ray morphologies (*e.g.* M86, Forman *et al.* 1979) and the soft spectrum typical of hot thin plasma (Trinchieri *et al.* 1986; Forman *et al.* 1985; Kim *et al.* 1992b). The X-ray data on the lower luminosity early type galaxies, although not of good enough statistics to give detailed information on each individual object, suggest that as a class their X-ray emission is probably dominated by mechanisms similar to those of the late types, based on the similarities between the X-ray luminosities and the spectral characteristics of the two classes. Individual sources, most likely low mass binaries, are therefore thought to be responsible for the observed X-ray properties.

The detection (or lack of) of a hot gaseous component in normal galaxies based on the *Einstein* observatory data needed to be confirmed with better quality data, and its properties needed to be further explored and more firmly established. The ROSAT satellite provides excellent capabilities to pursue these issues: more sensitive and more targeted observations could be planned, so that specific candidates could be observed for the time necessary to obtain statistically sound data, more objects could be observed to build a larger database, and this could all be done with instruments with a softer energy response, better suited for studies of the soft X-ray emission expected from hot gas. Some of the results obtained from these observations have been analyzed and reported in the literature already: besides a general confirmation of the *Einstein* results, these new observations have also added interesting information on the properties of the gaseous component in normal galaxies. A few of the observational results from ROSAT are given below.

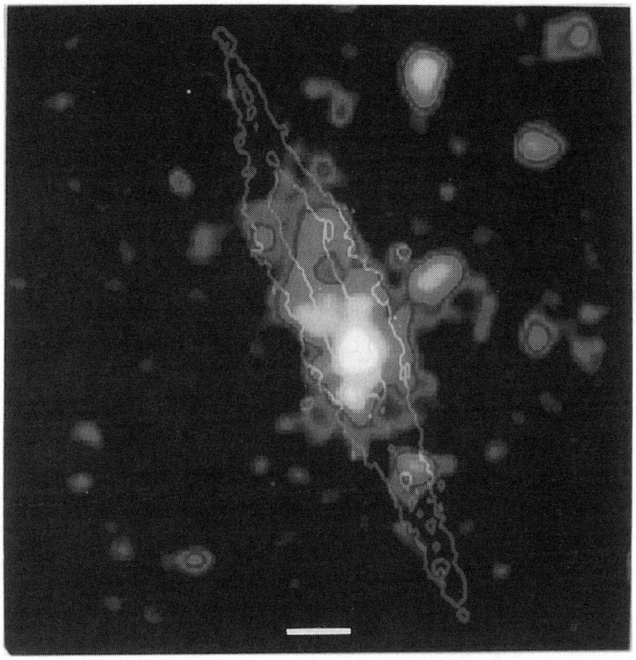

Figure 1. X-ray image of NGC 891 superposed onto the Hα isointensity contours (from Bregman & Pildis 1994)

2. Hot Gas in Late Type Galaxies

NGC 891 is one of the first galaxies observed by ROSAT for which a soft component, attributed to a thick disk of hot gas, is reported (Bregman & Pildis 1994). This galaxy was never observed with the *Einstein* Observatory so no comparison is possible. Fig. 1 shows the X-ray image of NGC 891 superposed on isointensity contours from the Hα data. There is clearly a good correlation between the X-ray and Hα extents, and also with the radio continuum halo. The X-ray emission is due to two main components: one, with a temperature of $kT_X > 2$ keV (basically outside the energy range of ROSAT, so it could not be measured except for the determination of a lower limit), could be due to the population of X-ray binaries found in all late type galaxies; and indeed one very bright source coincident with SN1986J plus two fainter ones are detected in the disk. The second component has a typical temperature of $kT_X \sim 0.3$ keV and a luminosity $L_X \sim 4.4 \times 10^{39}$ erg s^{-1}, and could be due to hot gas in the disk of the galaxy. The gas extent is of ~ 5 kpc in the direction perpendicular to the galaxy's plane and > 6 kpc along the plane, and has a density of $\sim 2 \times 10^{-3}$ cm^{-3}, for a total mass of $M \sim 1 \times 10^8 M_\odot$ (filling factor of unity). The gas is most likely bound to the galaxy: its temperature is lower than the escape temperature for this system, and the mass loss rate, if the gas were in a wind, compared to the replenishment rate from stars, would free this inner part of the galaxy from the galaxy in less than 10^9 years.

A jet is also observed that extends to the N-NW of the central region, also prominent in the soft band, with a luminosity of $L_X \sim 5 \times 10^{38}$ erg s−1. The nature of this feature is as yet unclear. A background origin is not ruled out. If, however, it is associated with NGC 891, it could represent emission related to a starburst event that expels gas out to $\sim$ 6 kpc from the galaxy's center.

The discovery of hot gas in a late type galaxy raises the question of whether this feature is common in these objects or whether NGC 891 is a special case. At the moment there are not enough data that would allow us to establish how common this component is in what kind of late type galaxies. It is, however, fair to say that while NGC 891 is not an isolated case, hot gas has not been detected in *all* of the late type galaxies observed. Let's look at some examples for different kinds of galaxies.

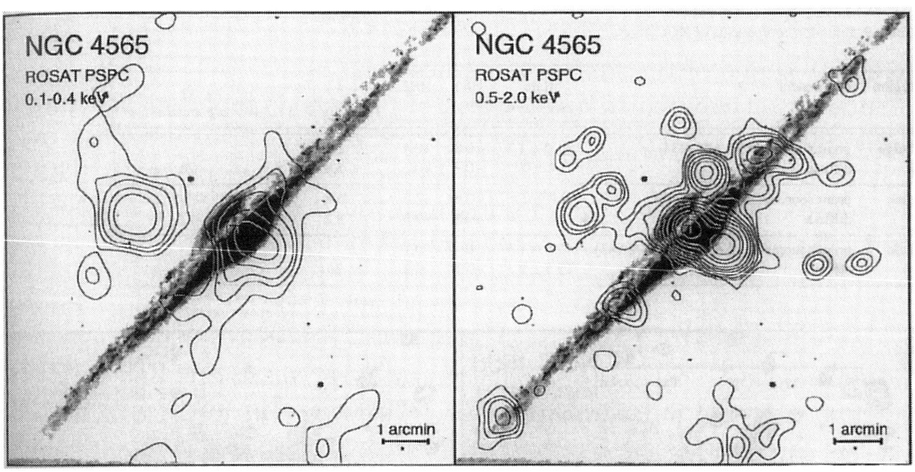

Figure 2. X-ray map of NGC 4565 superposed on an optical image (from Vogler *et al.* 1996).

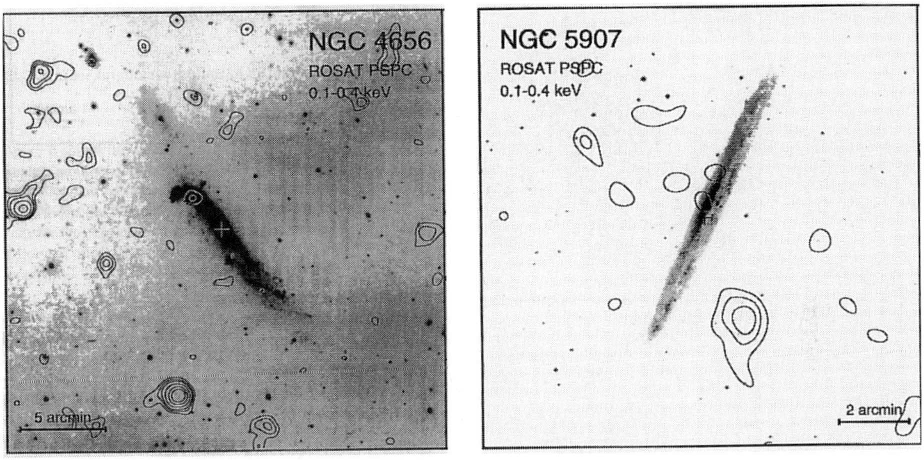

Figure 3. X-ray contours of NGC 4656 and NGC 5907 superposed on their optical images (from Vogler *et al.* 1996).

2.0.1. *Edge-on galaxies*

Besides NGC 891, other edge-on galaxies have been observed with ROSAT, and in some cases targets have been selected to obtain the most favorable conditions for the detection of the "soft" emission typical of a hot gaseous component.

A diffuse emission is detected outside the plane of NGC 4565 (Vogler et al 1996), shown in Fig. 2, to a height of $> 2'$. The emission to the NE is observed only in the soft energy range (0.1-0.4 keV), with a temperature of $T_X \sim 2 \times 10^6$ K. The X-ray luminosity is $L_X \sim 3 \times 10^{39}$ erg s^{-1} in the soft band, compared to a luminosity in the point source component of $L_X \sim 7 \times 10^{39}$ erg s^{-1}.

Some diffuse emission is also detected in the bulge and in the disk of this galaxy, but a crude estimate of its spectral characteristics indicates a relatively hard spectrum, which would associate this with emission from unresolved, lower luminosity individual sources, similar to the brighter ones individually detected.

In the same paper, Vogler *et al.* (1996) also report two examples of edge-on galaxies with no detection of a hot halo. In Fig. 3, the X-ray images of NGC 4656 and of NGC 5907 are presented, superposed on optical images. While it is evident in both cases that there are several individual sources, they are confined to the plane of the galaxy, and no indication for diffuse emission in the halos of these galaxies is found, with 2 σ upper limits down to a factor of 10 lower than the emission observed in NGC 4565.

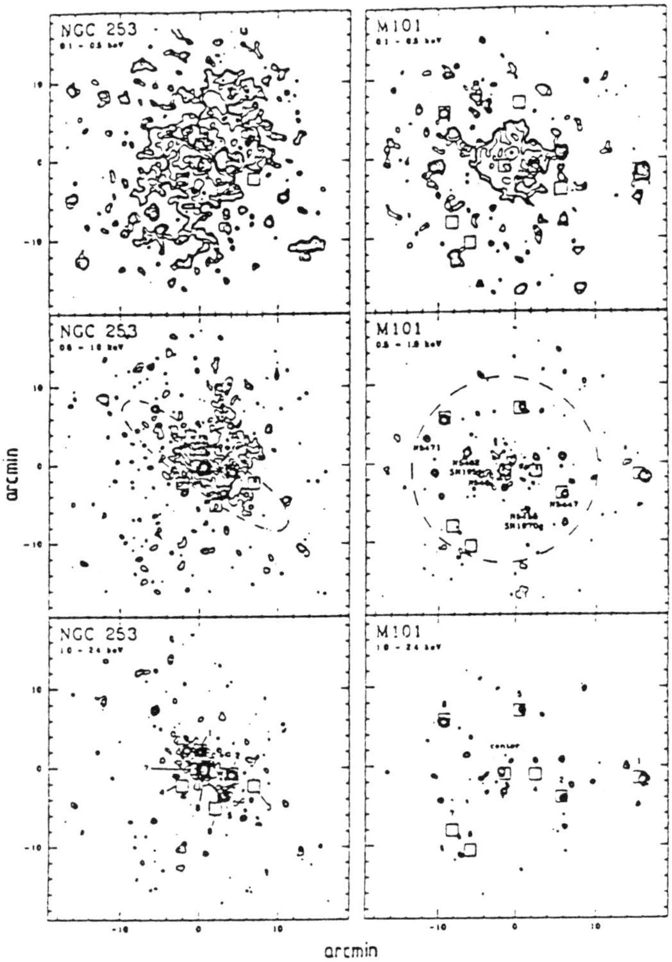

Figure 4. Isointensity X-ray maps of NGC 253 (left) and M101 (right) in three energy bands (softer on top, harder on the bottom). The top two panels show the diffuse soft emission discussed in the text (from Pietsch 1993).

2.0.2. *Face-on galaxies*

The *Einstein* and ROSAT observations of the face-on galaxy M101 have detected several point sources in the plane of this galaxy (Trinchieri *et al.* 1990; Snowden & Pietsch 1995). However, while no diffuse emission was found in the *Einstein* observations (McCammon & Sanders 1984), this component is detected from the inner part of the disk in the ROSAT data (Fig. 4). This emission is more prominent in the soft energy range (0.1-0.4 keV), appears to be peaked at the galaxy's center and could be due to a million degree gas with luminosity in the range $L_X \sim 10^{40-41}$ erg s^{-1}. Since this is a face-on galaxy, it is not clear whether this component is confined within the galaxy's disk or is located above it as it is observed in more edge-on galaxies (*i.e.* NGC 4565). A halo origin, however, is suggested by the comparison with other galaxies, including the Milky Way. In NGC 1566, a southern spiral, Ehle *et al.* (1996) find extended X-ray emission coinciding with the radio continuum emission. The X-ray luminosity of the extended component is $L_X = 1.4 \times 10^{40}$ erg s^{-1}. There are a few point sources besides the bright nucleus (section 2.0.4) coinciding with star forming regions in the disk. The late type galaxy NGC 4559 (Vogler *et al.* 1997) has similar features: an extended component ($L_X = 1.5 \times 10^{39}$ erg s^{-1}) and a few individual point sources in the disk. The point sources probably are associated with star forming regions and supernova remnants. This picture does not change significantly for very late type galaxies. The irregular galaxy NGC 4449 (Vogler & Pietsch 1997) also shows an extended component in addittion to the point sources in the disk. The extended component is less luminous ($L_X = 10^{39}$ erg s^{-1}) than the one in NGC 1566 and could very well arise from 3×10^6 K gas.

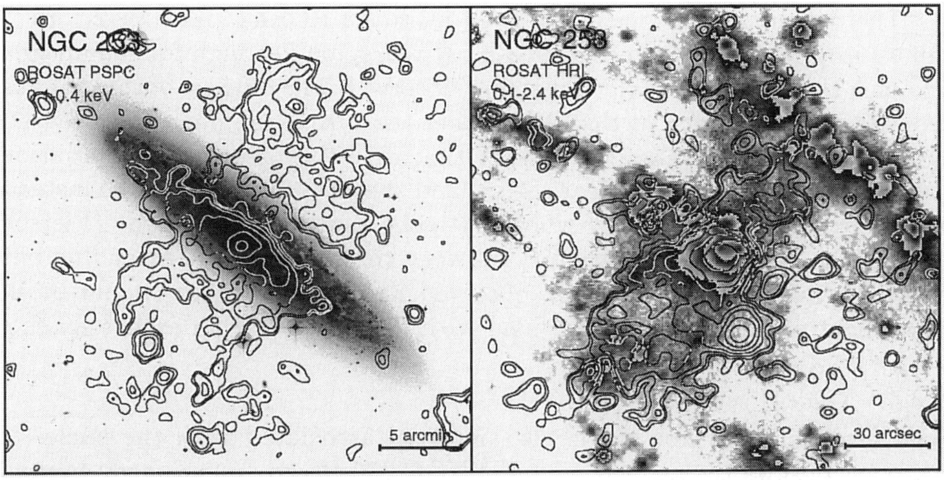

Figure 5. X-ray contours of the ROSAT PSPC and HRI data of NGC 253 in the soft energy band superposed onto an optical image of the galaxy (Courtesy of W. Pietsch & A. Vogler).

2.0.3. *Active/starburst galaxies*

One of the more spectacular examples of hot gas outside of the plane of spiral galaxies is offered by the image of NGC 253. The *Einstein* data had already shown a plume of emission extending out from the starburst nucleus, more prominent on the northern side (Fabbiano & Trinchieri 1987; Fabbiano 1989). ROSAT PSPC data have not only confirmed the presence of this extraplanar emission, but have added new interesting details (Pietsch 1993). In the very soft energy (0.1-0.5 keV) X-ray map (Fig. 4 and Fig. 5) there is evidence of unresolved emission from a large area, much less peaked towards the galaxy's center than the emission at harder energies, and more extended in the direction perpendicular to the galaxy's plane ($> 13'$ in both directions perpendicular to the plane, or $\sim$ 11 kpc). A hot ($T_X \sim 2 \times 10^6$ K) gas filling the halo of NGC 253 could well explain the observations.

A clear feature of this halo is that its X-ray emission is very inhomogeneous. To the South-East, the gas seems to flow out from the entire disk, not just from the nuclear region, and in fact reaches out further from the edges than from the central area. The emission to the North-West seems to extend more evenly from the entire disk, but a discontinuity (a $\sim 2' - 3'$ gap) is seen close to the disk edge (Fig. 5). This feature can be successfully explained by absorption by the intervening disk: the North-West side closer to us obscures the outflow perpendicular to the plane in the direction away from us. Only when far enough away from the disk, can the emission be seen coming out from behind it. An intervening absorbing column of only $N_H \geq 10^{20}$ cm^{-2} is required, very reasonable for a normal galactic disk.

The density of the plasma derived from the ROSAT data ($\sim 2 \times 10^{-3}$ cm^{-3}) implies cooling times of $\sim 2 \times 10^8$ years, long enough for the plasma to reach the distance at which it is now observed. The outflow of this hot gas from the whole length of the disk, rather than from the nuclear starburst, has raised questions about its association with the present star formation activity going on in the nuclear region, and suggests that it might be instead linked to some past activity whose optical signature is now gone (Pietsch 1993). The observed discontinuity between the X-ray plume extending out from the nucleus and the halo, observed at high resolution, would be in agreement with the halo being filled with the hot gas at an earlier epoch.

2.0.4. *Nuclear regions*

In a few instances hot gas is also detected associated with the nuclei of galaxies, besides the nuclear region of NGC 253 already mentioned. A small plume extends out from the nucleus of NGC 247, an edge-on galaxy that shows a few bright X-ray sources but no other extended emission in the plane (Mackie *et al.* 1994). Emission from hot outflowing gas is also reported from the nucleus of M51 (Ehle *et al.* 1995), in coincidence with the radio jet. A 400 pc diameter diffuse emission is detected in the central region of IC 342 (Bregman *et al.* 1993), approximately coincident with both the radio continuum and the Hα emissions. A central point source, contribution roughly equal amount to the diffuse emission is expected to be present as well, giving rise to a total luminosity in the region of $L_X \sim 1.4 \times 10^{39}$ erg s^{-1}. A recent starburst phenomenon, with age $\leq 10^5$ years, is probably responsible for the presence of the hot gas in this galaxy, with an estimated average density of 0.2 cm^{-3}, for an assumed temperature of 1 keV and filling factor of unity, and a pressure comparable to the pressure in the molecular gas also present in the region. The face on Seyfert galaxy NGC 1566 has a bright central source of $L_X = 10^{41}$ erg s^{-1} with a power law spectrum of index 2.3 (Ehle *et al.* , 1996) which appears typical for Seyfert galaxies (Walter and Fink, 1993). Junkes *et al.* (1995) find a strong unresolved X-ray source in the starburst nucleus of NGC 1808. This source has a similar luminosity of $L_X = 1.4 \times 10^{41}$ erg s^{-1} probably from hot (6×10^6 K) gas. NGC 1808 does not show extended X-ray emission as does NGC 1566. The nuclear X-ray source in NGC 1808 could be powered by supernova remnants or hot bubbles provided there is a massive star formation rate of $\sim 10 M_\odot \mathrm{yr}^{-1}$, in agreement with estimates from the FIR properties.

2.1. IN SUMMARY: WHAT ARE THE PROPERTIES OF HOT GAS IN SPIRAL GALAXIES?

As a result of the ROSAT data, hot gas has been found in *some* of the galaxies observed. It is too early, however, to establish what the general properties of the hot interstellar medium in late type galaxies are, because of the limited number of objects studied and also because the characteristics and the morphology vary.

Hot gas is found in the form of a thick disk, in the plane of face-on objects, in the halo region either as a plume extending from the nuclear region or as an asymmetric halo extending on both sides of the galaxy and originating from the whole disk. Sometimes only the nuclear region shows evidence of hot gas. At the same time, no evidence of diffuse emission is seen in some other galaxies down to levels comparable to the detections.

The lack of detections could either be due to some intrinsic differences in the properties of these galaxies, that cannot heat their interstellar medium or cannot retain the hot gas, or to differences in the characteristics of the Inter Stellar Medium (ISM) itself, that could be for example too cool ($T_X < 10^6$ K) to be detected by ROSAT, or could be absorbed by intervening cold gas in the galaxy. Hot halos could also be the signature of starburst activity: recent events are still detectable at other wavelengths, for past events the hot gas represents the only visible remnant, still detectable because of its relatively longer cooling time.

When detected, the gas has luminosities of $L_X \sim 2 - 5 \times 10^{39}$ erg s^{-1}, with temperatures of $T_X \sim 2 - 5 \times 10^6$ K, well matched to the energy response of ROSAT, and also just below the capability of the *Einstein* instruments. Cooling times for the gas are in the range $10^7 - 10^8$ years and the gas masses are $M_{gas} \sim 10^7 - 10^8\ M_\odot$.

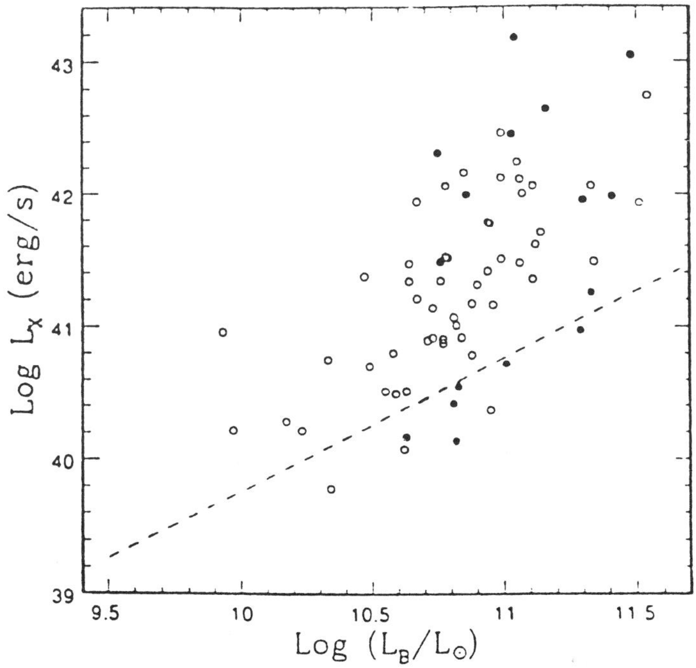

Figure 6. L_X-L_B diagram for the early type galaxies detected by *Einstein* (data from Fabbiano *et al.* 1992). The diagonal line identifies the expected contribution from the evolved stellar population, as observed in the late type sample. Filled circles identify galaxies for which ROSAT data have been analyzed and are discussed in the text.

3. Hot Gas in Early Type Galaxies

Einstein Observatory data on early type galaxies have shown that a correlation exists between their X-ray and optical luminosities. As illustrated in the diagram in Fig. 6, the correlation is, however, much steeper than linear ($L_X \sim L_B^{1.7-2}$ Trinchieri & Fabbiano 1985) and the scatter about the best fit slope is much larger than what can be expected from measurement uncertainties: the ratios of L_X/L_B could vary as much as a factor of 100, for optical luminosities of $L_B \geq 10^{11}\ L_\odot$. This diagram has been interpreted as evidence that several mechanism are at play in different regions, from the hot gas dominating the emission at the bright X-ray luminosity end to the evolved stellar population emission at the faint end. The balance between the gravitational potential of the galaxies and the energy inputs would act to suppress winds in the bright objects, and inversely would sweep the galaxies free of gas, through galactic winds, for the X-ray faint galaxies (see Ciotti *et al.* 1991).

Strong supporting evidence for this interpretation comes from the results on the spectral characteristics of the emission. Although galaxies could not be studied individually, due to the limited statistics provided by the *Einstein* observations, especially at the faint X-ray luminosity end, an average spectrum was derived for different groups of L_X/L_B ratios (Kim *et al.* 1992a,b). A steepening of the average spectrum is found as the L_X/L_B ratios decrease (Fig. 7) in agreement with the increasing importance of the stellar population component in the X-ray emission. The spectra of the two lower panels are consistent with a $\sim$ 1 keV spectrum, while the upper right panel shows that the data are consistent with the harder average spectrum of the late type galaxies, whose emission is dominated by individual, mostly binary, sources. In the lowest L_X/L_B bin (upper left panel) an excess in the soft energy channels is seen, that can be explained with a second, very soft ($kT_X \sim 0.2$ keV), component added to the hard one.

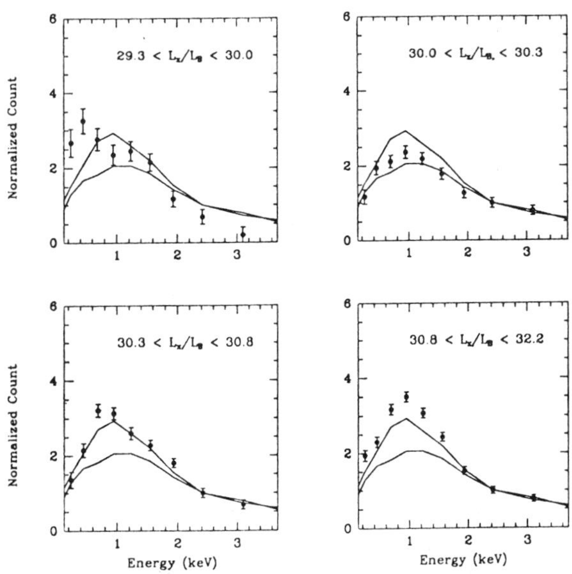

Figure 7. Average spectra derived for the co-added IPC data of early type galaxies divided in four bins of L_X/L_B ratios. The two solid lines represent the co-added spectra of the entire early (upper line, $kT_X \sim 1$ keV) and late (lower line, $kT_X \sim 5$ keV) type galaxy samples, arbitrarily normalized.

Objects of different L_X/L_B ratios have now been observed with ROSAT

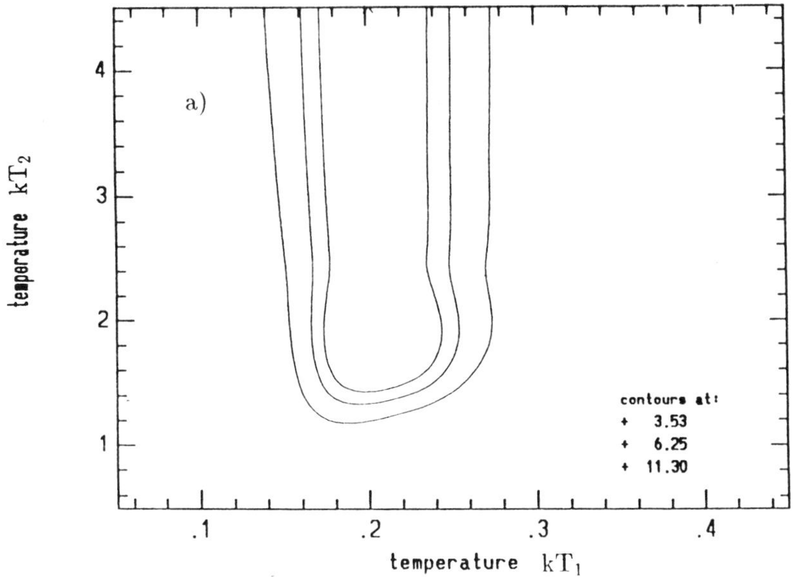

Figure 8. The confidence region for the two temperature in the spectrum of NGC 4365 (from Fabbiano, Kim & Trinchieri 1994).

(see Fig. 6). The analysis of the data for galaxies in the lowest L_X/L_B ratio group confirms the average spectral results of the *Einstein* data: a very soft component, with $kT_X \sim 0.1$-0.3 keV is observed (Fig. 8), that accounts for $\sim 50\%$ of the total observed X-ray luminosity. The nature of this component is still not totally understood: if these galaxies do not retain a hot ISM, this emission could be due to the stellar population, such as Super Soft Sources, like those found in nearby galaxies (Griner *et al.* 1991; Kahabka *et al.* 1994) or stellar coronae (M stars; Schmitt *et al.* 1990). Simulations, however, have shown that these galaxies could retain some of their ISM in a much cooler phase than found in brighter objects (Pellegrini & Fabbiano 1994), which would account for this softer component.

At the other end of the L_X-L_B digram, hot gas dominates the emission. In the *Einstein* data, the hot gas has a luminosity $L_X \sim 10^{41} - 10^{43}$ erg s^{-1}, with a relatively regular, round X-ray morphology, roughly co-extensive with the stellar light, and with temperatures, averaged over the whole source, in the range $kT_X \sim 0.5 - 2$ keV. On the assumption of azimuthal symmetry, the surface brightness distribution has been approximated by King-type profiles of the kind:

$$\Sigma_X \propto [1 + (\frac{r}{r_c})^2]^{-\beta+0.5}$$

where r_c, the core radius, of the order of $\sim 10''-15''$, could only be measured for a very few objects observed at higher spatial resolution. The parameter β is ~ 0.5, so at large radii, $\Sigma_X \sim r^{-2}$, with relatively little scatter.

With these results, and the assumption that the gas is isothermal and spherically symmetric, intrinsic gas parameters have been derived: the gas density, distributed as $\rho \sim r^{-1.5}$, with central values of $\rho \sim 0.1 - 1$ cm^{-3} and $\rho \sim 10^{-4}$ cm^{-3} at the edges; cooling times shorter than a Hubble time over most of the galaxy extent; and total gas masses amounting to $\sim 10\%$ of the luminous mass in the system.

The detection of hot gas out to radii comparable to the optical size of these galaxies has been used to derive total gravitational masses in these systems, under a few simple assumptions that the gas is in hydrostatic equilibrium and is distributed with spherical symmetry (Fabricant & Gorenstein 1983). With the quality of the *Einstein* data however, the determination of the gravitational masses could only be obtained with factors of ~ 10 uncertainties, giving mass-to-light ratios $M/L \sim$ 10-80 $L_\odot/M_\odot$ (Trinchieri *et al.* 1986). While these values might indicate the presence of dark matter ($M/L \sim 6 - 10$ are expected from the luminous mass), they are still uncertain not only because of the limited statistics, but also because they depend on some relatively important and unproven assumptions, mostly the radial dependence of the temperature and the spectral model used to derive them (Trinchieri *et al.* 1977).

Several X-ray bright galaxies have now been observed with ROSAT. The statistical evidence is still based on only a few objects so that strong conclusions cannot be drawn yet. However, it is already evident that the picture is more complicated than *Einstein* data had led us to believe.

3.0.1. *Morphology of the gas*

The first evidence is the morphological appearance of the gas (some examples are shown in Fig. 9): while a few objects are indeed fairly regular and round (NGC 4649 and NGC 7626 for example) others present some peculiarity, from the elliptical shape of the isophotes of NGC 533, to the complex emission of NGC 4636, round and azimuthally symmetric in the centre, but more elongated towards the SW at R $\sim 6'$ and more pronounced to the North outside $R \sim 8' - 10'$ (incidentally, M87, at the center of the Virgo cluster, is ~ 3 Mpc away to the north; Trinchieri *et al.* 1994). More extreme cases are the tail seen to the SW of NGC 7619, that extends for $\sim 10'$ away from the galaxy's center, and the extent of NGC 2563 relative to the optical galaxy.

The striking difference between the X-ray morphology of NGC 7619 and of NGC 7626, both in the same Pegasus I group, poses a question as to how much the environment influences the X-ray morphology. The

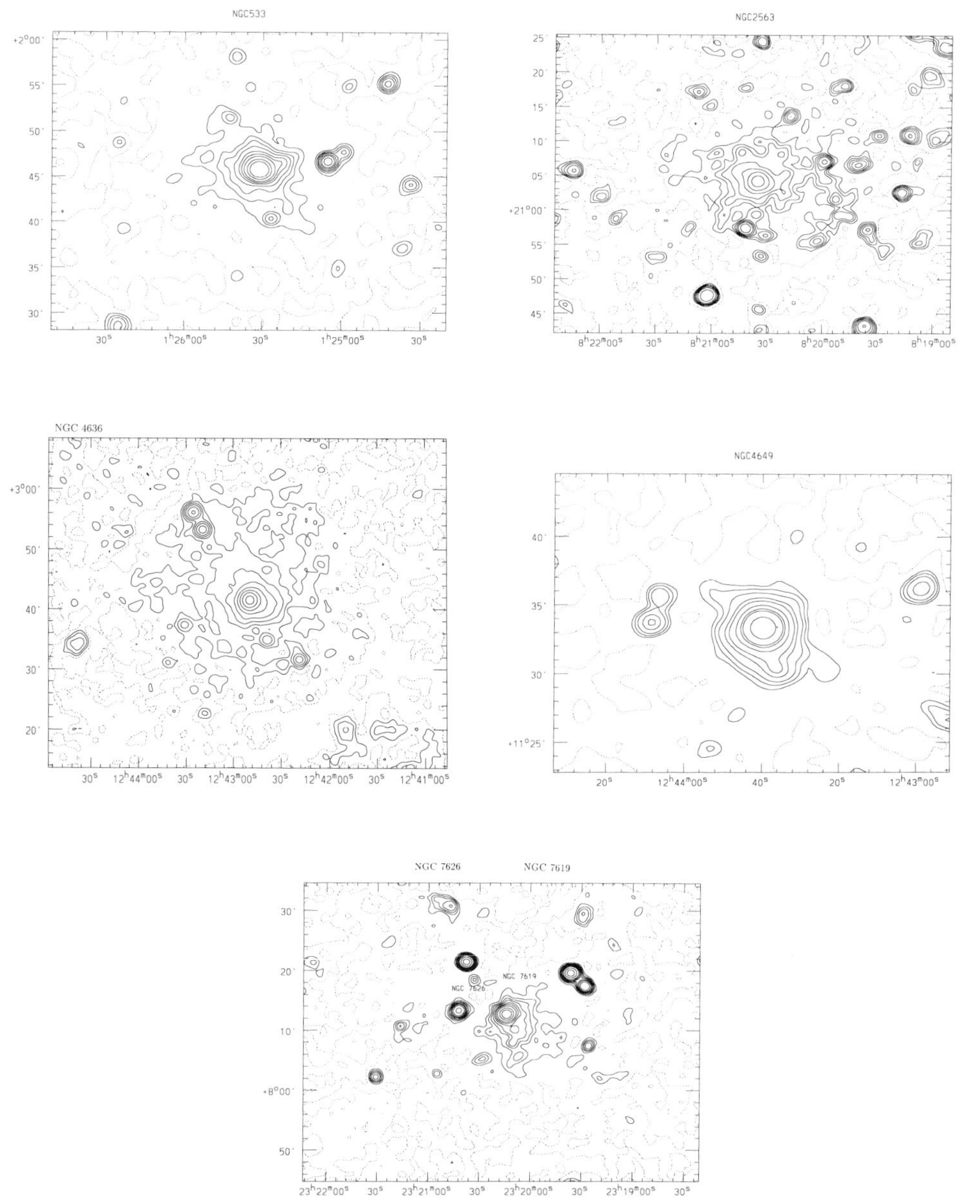

Figure 9. X-ray isointensity contour maps of the 6 early type galaxies discussed in the text. The crosses in the maps of NGC 7619/7626 and NGC 2563 indicate the position of galaxies in the group.

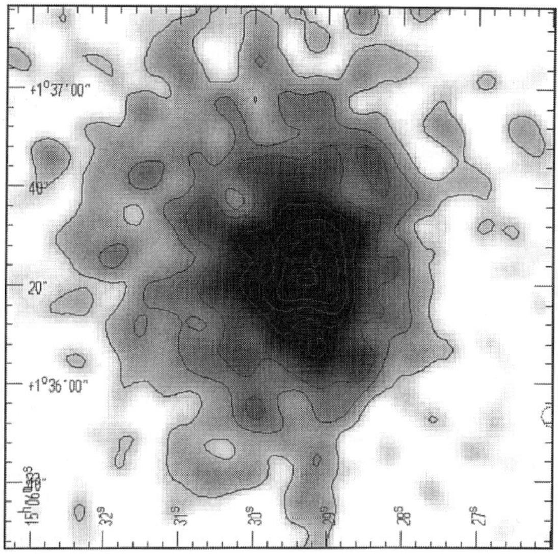

Figure 10. X-ray image of NGC 5846 at high spatial resolution.

examples above are found in somewhat different environments, from the fairly isolated NGC 533 and NGC 4636 (although this latter is a member of the Virgo cluster, $> 10°$ away from the cluster center), to NGC 4649 in the Virgo cluster, relatively close to M87. The other important question is where do galaxies end and other components (*i.e.* group or cluster) start to dominate. In the map of NGC 2563, the presence of a group (subgroup A in the Cancer cluster, Bothun *et al.* . 1983) could for example explain the large extent, and indeed several peaks of emission seem to be spatially coincident with other galaxies in the group.

The gas morphology at small scales is also quite different than what was suggested from the *Einstein* data. The X-ray contours of the central regions of a few galaxies observed at higher resolution indicate that the emission could be highly irregular, distorted and patchy, like in the image of NGC 5846 shown in Fig. 10. Clumpiness is also observed at large radii as in the case of NGC 507 (Fig. 11, Kim & Fabbiano 1994). These clumps could indicate localized cooling of the hot gas both at small and at large radii, since there do not seem to be obvious optical counterparts on the Palomar Sky Survey that could explains them as interlopers. Moreover, in the NGC 507 case, these clumps seem to have a softer spectrum than the surrounding gas.

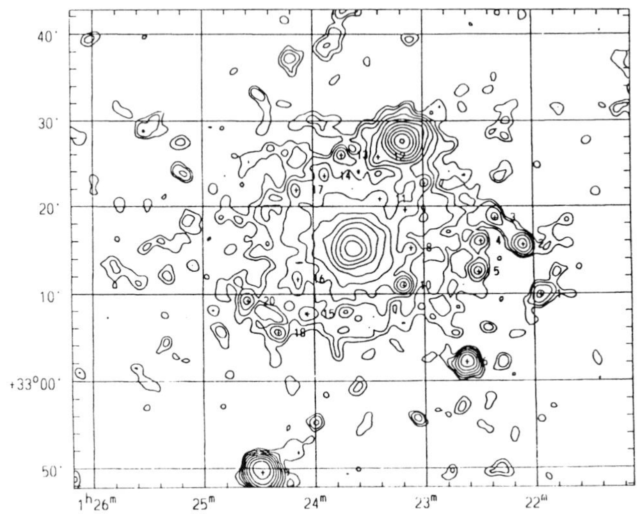

Figure 11. Contour map of the PSPC image of NGC 507 (NGC 499 is also in the field to the N-W). Numbers indicate the individual clumps detected (from Kim & Fabbiano 1994).

3.0.2. *Radial distribution of the gas*

The radial distribution of the emission also reflects the differences in the morphology, so that Σ_X can vary with radius from $\Sigma_X \propto r^{-1}$ (*e.g.* NGC 2563, Fig. 12, NGC 5044, David *et al.* 1994) to $\Sigma_X \propto r^{-3}$ (*e.g.* NGC 4649). This could very well reflect the fact that we are looking at different linear scales in different objects – in NGC 2563 and NGC 5044 we are probably observing mostly the group, while in NGC 4649 we are inside the stellar sphere. Moreover, while these profiles have been obtained assuming azimuthal symmetry, in some cases this is a poor approximation, and profiles at different angles or in different regions are significantly different, as in the case of NGC 533 and NGC 7619 (see Trinchieri 1995)

A further element that complicates matters is the possible presence of multiple components in the emission, as in the case of the tail in NGC 7619, and of the complex morphology of NGC 4636. The radial profile of NGC 4636 is shown in Fig. 13. In the left panel the profile of the whole emission is shown, with the background taken at large radii: a significant change in slope at $R \sim 6' - 8'$, roughly where the stellar light profile "ends" (see figure), can be clearly seen. If the change in slope is a sign that a different component is present (it could be either a background/foreground extended source; or emission from a group, although there is no optical evidence for galaxies of comparable size and distance nearby; or the Virgo cluster that extends out to here, $\geq$3 Mpc south of M87), then it might be necessary to subtract it from the galaxy's emission. In this case, the profile derived

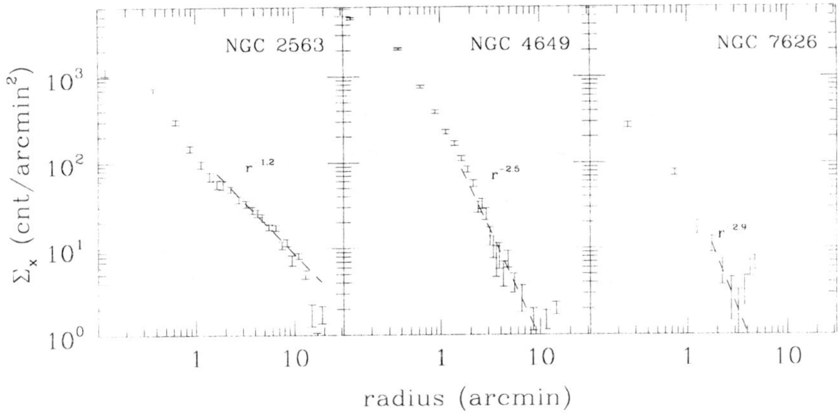

Figure 12. Azimuthally averaged surface brightness distribution of three of the early type galaxies discussed.

for this galaxy (right panel) also becomes rather steep, similar to that of NGC 4649.

Radial distributions of the stellar light are available for only a few galaxies, so a comprehensive comparison between optical and X-ray profiles will have to be done when more data are available. In the few cases examined so far, however, the two light distributions are very similar for most of the galaxy's extent (see Fig. 13; Trinchieri 1995).

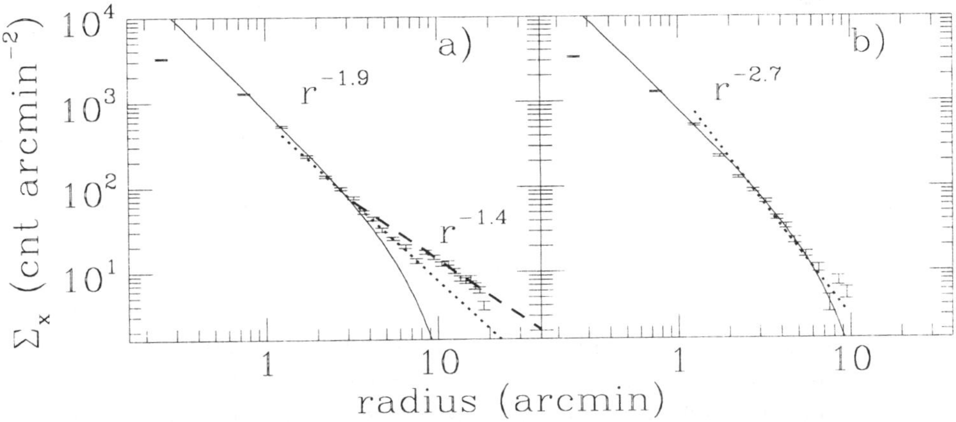

Figure 13. Azimuthally averaged surface brightness distribution of NGC 4636. In the left panel the background is taken at large radii, in the right one the emission outside 10′ is considered background. The solid line indicates the optical profile, the dashed lines are the power law fits to the X-ray profile

3.0.3. *X-ray spectra*

Radial distribution of the gas temperatures have now been determined for several objects, and in most cases indicate that the gas is not isothermal over the whole source: the central parts are significantly cooler than the outskirts. In some cases the cooling is confined to a small region, outside of which the gas tends to become more isothermal; in others it extends over most of the source (Fig. 14). The reason for this difference could be once again either due to the different linear scales explored in different objects, or to the presence of different components or both.

The errors in the determination of the temperature have now become quite small (see Fig. 14). These must, however, still be derived subject to the assumption of a —ixspectral model, which cannot be determined directly from the ROSAT data, since the results obtained from the fits with significantly different models applied to the same data give equally satisfactory results. Fig. 15 shows the data in the inner region of NGC 4636, which has relatively high statistics, compared to three different spectral models ('Raymond-Smith' plasma with low energy cut-off) to illustrate this ambiguity. With 100% cosmic abundances, good fits are obtained with multitemperature models, and possibly with a high intrinsic absorption. For a model with only one temperature, the metal abundance has to be very low. In all cases, all three models represent the data equally well, as shown by the distribution of the residuals (bottom panels) and by the best fit χ^2 values. Fortunately in all three cases there is one temperature (the one temperature in the 20% cosmic abundance model, and the dominant component in the other two) that can be determined well and is the same in all three models.

Clearly additional input is needed to determine which is the correct spectral model to use, since the ROSAT X-ray data alone cannot be a powerful discriminant. If the metal enrichment is close to the cosmic value, then the gas is probably inhomogeneous, *i.e.* at several temperatures at each radius. If, however, the gas is *very* metal-poor, than the temperature of the gas increases monotonically towards the outer parts of the galaxies, but we have to explain why gas that is shed from metal rich stars has such a different composition, or why it has been so strongly diluted. ASCA data, with their higher spectral resolution, should be important in helping us to discriminate between the spectral models, even thought they will be tested on the emission averaged over larger regions due to the poorer spatial resolution of its instrumentation. The results available so far would indeed suggest that the gas has a metal abundance significantly below the cosmic value (Awaki *et al.* 1994). However, even with ASCA data, caution should be taken in interpreting the data as coming from a gas of sub-solar abundance, since a multitemperature plasma can often be fit just as well

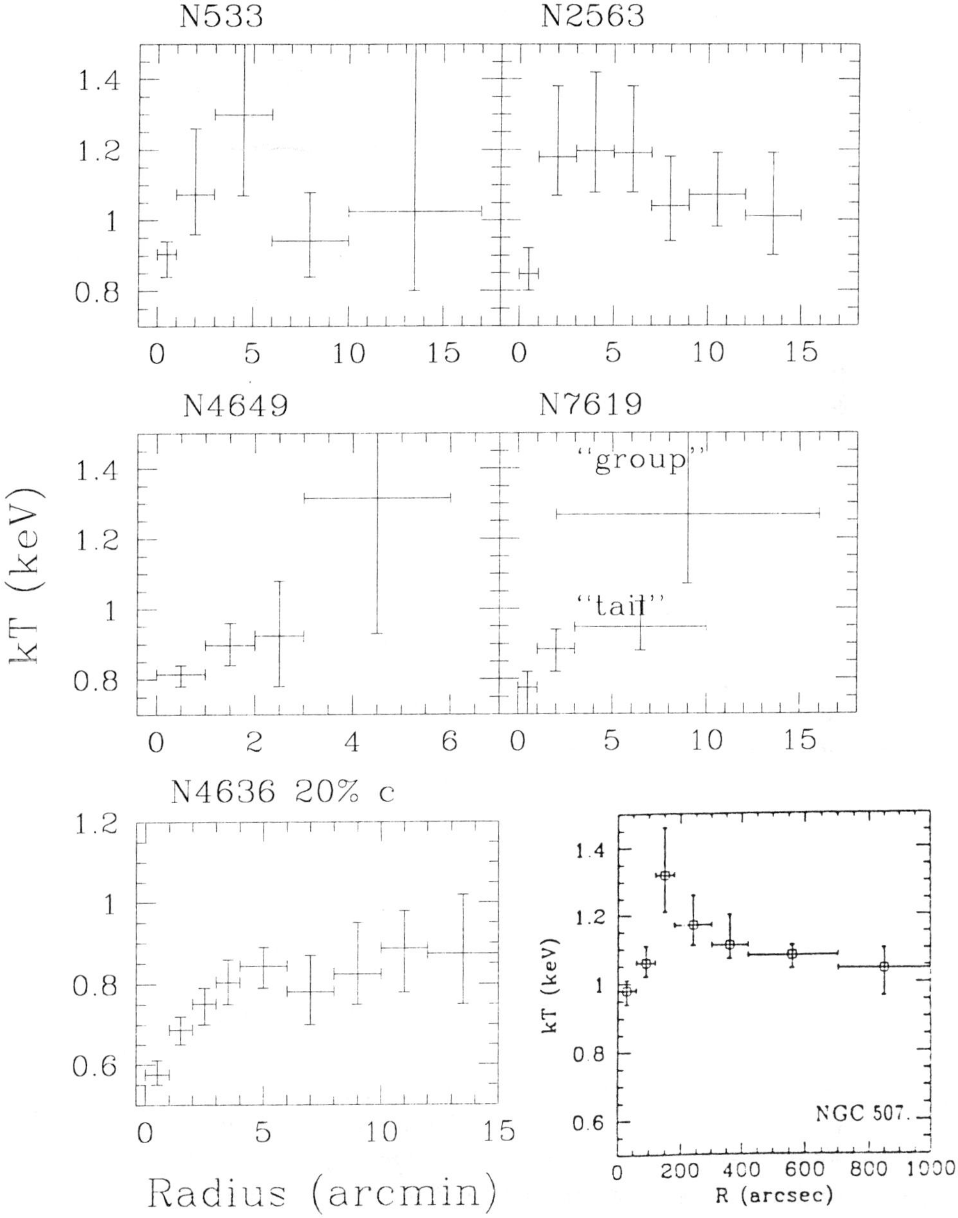

Figure 14. Distribution of the gas temperatures in different galaxies.

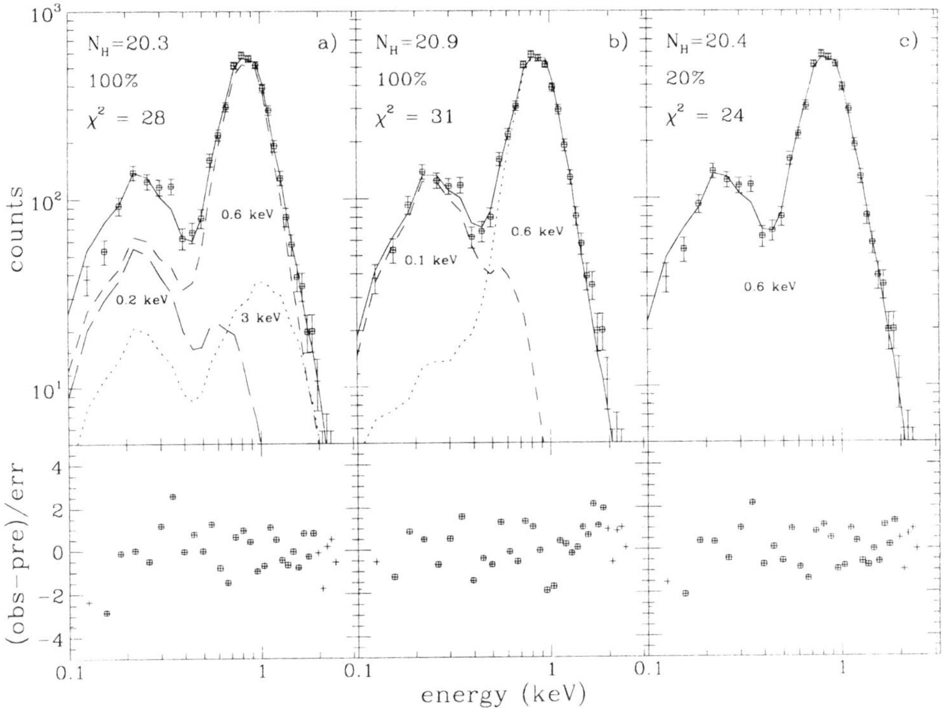

Figure 15. Spectral distribution of the counts in the inner region of NGC 4636 compared to the 3 different best fit thermal models: a) 3 temperatures, 100% cosmic abundance and N_{H} at the line-of-sight galactic value; b) 2 temperatures, 100% cosmic abundance and N_{H} much higher than galactic; c) 1 temperature, 20% cosmic abundance, N_{H} roughly at the galactic value.

to the data (Kim *et al.* 1997).

3.0.4. *Total gravitational masses*

Although the spectral characteristics of the hot ISM in early type galaxies cannot be measured unambiguously under the assumption of low metal abundance, a single temperature can be measured at each radius with small uncertainties. Temperature and density profiles can then be derived and used to estimate the total gravitational mass M_{T} in the system. With the further assumptions of spherical symmetry (which will, however, not be correct in some cases) and hydrostatic equilibrium, M_{T} is given by:

$$M_{\mathrm{T}}(R) = -\frac{\mathrm{k}T_{\mathrm{X}}}{G\mu m_{\mathrm{H}}}\left(\frac{\mathrm{dlog}\rho}{\mathrm{dlog}R} + \frac{\mathrm{dlog}T_{\mathrm{X}}}{\mathrm{dlog}R}\right)R$$

(see Fabricant and Gorenstein 1983), and depends therefore on quantities all of which are measurable from the X-ray data. With the reduced error bars on each of the items in the equation, the determination of $M_{\rm T}$ is also greatly improved over previous estimates.

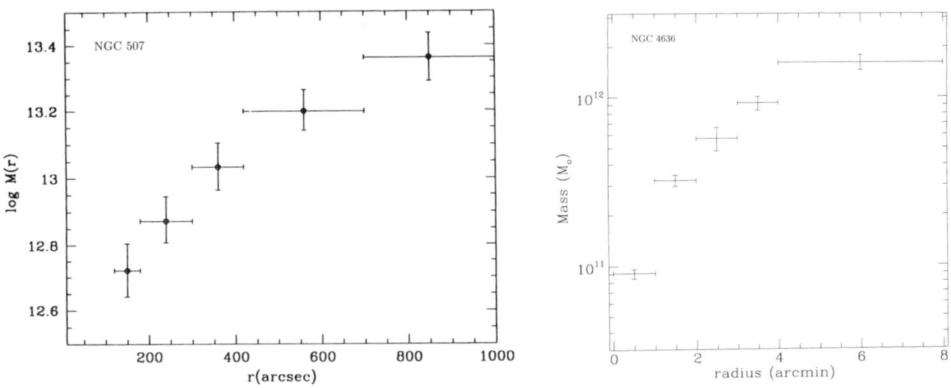

Figure 16. Total gravitational mass distribution for NGC 507 (left panel) and NGC 4636 (right panel).

Fig. 16 illustrates the radial behaviour of $M_{\rm T}$ for NGC 4636 and NGC 507 (Kim & Fabbiano 1994). At the outermost radius the mass-to-light ratio $M/L \sim 28$–45 for NGC 4636, and $M/L \sim 100$ for NGC 507. For comparison, with *Einstein* data, the M/L of NGC 4636 was measured in the range 15->100 (Trinchieri *et al.* 1986).

3.1. SUMMARY OF THE PROPERTIES OF THE HOT GAS IN EARLY TYPE GALAXIES

From the few examples examined so far it is already clear that hot gas may exist in all early type galaxies, but with different characteristics. If it exists in the low X-ray luminosity objects, its temperature must be very low, compared to brighter objects. In these, the morphology could be heavily distorted, possibly due to external pressure (from the group or the cluster gas), and could present inhomogeneities, clumps and asymmetries at all scales.

The temperature of the gas can now be measured with much greater accuracy than previously, but it still depends on the assumption of the metal composition of the gas. There is a general tendency for temperatures to increase towards larger radii, on scales comparable to the galaxy's size. Good spectral data and more sophisticated models are still needed to determine the gas metal abundance, a crucial parameter for the understanding of the evolution of early type galaxies. High resolution sensitive spatial data, coupled with the spectral information, are also needed to separate the contribution of different components, *i.e.* emission from the group vs. emission from the galaxy, an issue that will become even more crucial as larger and more distant samples become available.

I would like to thank G. Fabbiano, D-W. Kim, L. Noris, W. Pietsch, and A. Vogler, who have all been very helpful and have provided me with excellent material from their work to prepare for this chapter.

References

Awaki, H, *et al.* 1994 *P.A.S.P.* , 46 L65.
Bothun, G. *et al.* . 1983, *Ap.J.* , 268, 47.
Bregman, J.N., Cox, C.V., Tomisaka, K. 1993, *Ap.J.* , 415, L79.
Bregman, J.N. and Pildis, R.A. 1994, *Ap.J.* , 420, 570.
Ciotti, L., D'Ercole, A., Pellegrini, S., and Renzini, A. 1991, *Ap.J.* , 376, 380.
David, L. *et al.* 1994 *Ap.J.* , 428, 544.
Ehle, M., Pietsch, W. and Beck R. 1995, *Ast.Ap.* 306, 73.
Ehle, M., Beck, R., Haynes, R. F., Vogler, A., Pietsch, W., Elmouttie, M. and Ryder, S. 1996, *Ast.Ap.* , 306, 73.
Fabbiano, G., 1988 *Ap.J.* , 330 672.
Fabbiano, G., 1989, *Ann.Rev.Ast.Ap.* . 27, 87.
Fabbiano, G., Heckman, T., and Keel, W.C. 1990, *Ap.J.* , 355, 442.
Fabbiano, G. and Trinchieri, G. 1984, *Ap.J.* , 286, 491.
Fabbiano, G. and Trinchieri, G. 1987, *Ap.J.* , 315, 46.
Fabbiano, G., Kim, D-W. and Trinchieri, G. 1992 *Ap.J.Suppl.* , 80, 531.
Fabricant, D. & Gorenstein, P. 1983 *Ap.J.* , 267, 535.
Forman, W. *et al.* 1979 *Ap.J.* , 234, L27.
Forman, W., Jones, C., and Tucker, W. 1985 *Ap.J.* , 293, 102.
Griner, J. Hasinger, G. & Kahabka, P. 1991 *Ast.Ap.* , 246, L17.
Junkes, N., Zinnecker, H., Hensler, G., Dahlem, M. and Pietsch, W. 1995, *Ast.Ap.* , 294, 8.
Kahabka, P., Pietsch. W. and Hasinger, G. 1994 *Ast.Ap.* , 288, 538.
Kim, D-W., Fabbiano, G. and Trinchieri, G. 1992a *Ap.J.Suppl.* , 80, 645.
Kim, D-W., Fabbiano, G. and Trinchieri, G. 1992b *Ap.J.* , 393, 134.
Kim, D-W., & Fabbiano, G. 1994, *Ap.J.* , 429, 94.
Kim, D-W. *et al.* 1997, in preparation
McCammon, D, Sanders, W.T 1984, *Ap.J.* , 287, 167.
Mackie, G. *et al.* , 1994 in preparation
Pellegrini, S. and Fabbiano, G. 1994 *Ap.J.* , 429, 105.
Pietsch, W. 1993, 'Panchromatic View of Galaxies', G. Hensler, Ch. Theis, J. Gallagher (eds). p. 137.
Schmitt J. *et al.* 1990 *Ap.J.* , 365, 704.

Snowden, S.L. Pietsch, W. 1995, *Ap.J.* , 452, 627.
Trinchieri, G, 1995, in *The physics of Interstellar and Intergalactic Medium*, A. Ferrara, C. Heiles, C.F. McKee, P. Shapiro (eds). ASP Conf. nr. 80, p. 409.
Trinchieri, G., Fabbiano, G. 1985, *Ap.J.* , 296, 447.
Trinchieri, G., Fabbiano, G., and Canizares, C. 1986, *Ap.J.* , 310, 637.
Trinchieri, G., Fabbiano, G. and Peres, G. 1988, *Ap.J.* , 325, 531.
Trinchieri, G., Fabbiano, G. and Romaine S. 1990, *Ap.J.* , 356, 110.
Trinchieri, G. *et al.* 1994, *Ap.J.* , 428, 555.
Trinchieri, G., Fabbiano, G. and Kim, D-W. 1997, *Ast.Ap.* , 318, 361.
Vogler, A., Pietsch, W. and Kahabka, P. 1996, *Ast.Ap.* , 305, 74.
Vogler, A. and Pietsch, W. 1996, *Ast.Ap.* , 311, 35.
Vogler, A. and Pietsch, W. 1997, *Ast.Ap.* , 319, 459.
Vogler, A., Pietsch, W. and Bertoldi, F. 1997, *Ast.Ap.* , 318, 768.
Watson, M.G. Stanger,V. and Griffiths, R. 1994, *Ap.J.* 286, 144.
Walter, and Fink, H.H. 1993, *Ast.Ap.* ,

ABUNDANCE GRADIENTS IN SPIRAL GALAXIES

M. B. VILA-COSTAS
Department of Physics and Astronomy
University of Wales College of Cardiff
P.O. Box 913
Cardiff CF1 3TH
U.K.

1. Introduction

The study of H II regions in the disks of spiral galaxies has given us important information about their physical conditions. In particular the study of their chemical abundances soon showed the existence of negative gradients with higher abundance towards the centre of a galaxy (e.g. Pagel and Edmunds 1981; McCall 1982; Díaz 1989; Dinerstein 1990).

The origin of these gradients is not clear yet, but there are various theoretical models that can produce them with either a radial variation in the star formation rate and gas fraction (Pagel and Edmunds 1981; Garnett and Shields 1987), a variable yield due to variations in the initial mass function (IMF) (Güsten and Mezger 1982), a variation in the ratio of star formation rate to rate of inflow of unprocessed gas (Díaz and Tosi 1984) or radial flows of gas relative to stars (Mayor and Vigroux 1981; Lacey and Fall 1985; Pitts and Tayler 1989; Clarke 1989; Sommer-Larsen and Yoshii 1989). More observational data is necessary to allow us to decide which of these effects is dominant.

Until recently the main observational result found was the correlation between the oxygen abundance gradient and the surface density (Edmunds and Pagel 1984; Garnett and Shields 1987), which was present in spirals as well as in elliptical and dwarf galaxies (Edmunds and Phillipps 1989; Phillipps, Edmunds and Davies 1990). There were indications of other possible correlations between the oxygen abundance gradient and the Hubble type, the luminosity or absolute magnitude of the parent galaxy (Garnett and Shields 1987), or the cluster environment (Shields, Skillman and Kenni-

J. M. van der Hulst (ed.), The Interstellar Medium in Galaxies, 153–170.

cutt 1991). However it has not been until recent studies that more positive results have been found.

In this chapter I will review the latest results in relation with the oxygen abundance in spiral galaxies. They include the studies by Vila- Costas and Edmunds (1992) (VE92), with a sample of 30 galaxies from the literature; Oey and Kennicutt (1993) (OK93), with 15 new Sa, Sb galaxies which increased dramatically the sample of these early types; Zaritsky, Kennicutt and Huchra (1994) (ZKU94), with 14 new galaxies and their compilation of a total sample of 39 galaxies with more than 5 H II regions and Martin and Roy (1994, 1995), with their study of barred galaxies, adding 3 new ones for a total sample of 24 galaxies with more than 10 H II regions (5 barred, 8 non-barred and 11 mixed types).

These review will not discuss the abundances of other elements, e.g. nitrogen (Vila-Costas and Edmunds 1993).

2. Background Information

I would like to remind the readers of certain factors to be taken into account when determining the oxygen abundances in spiral galaxies. (see Dinerstein (1990) for a more detailed account).

2.1. ABUNDANCE DETERMINATION

2.1.1. *Direct method*

The oxygen abundances can be obtained through the direct method which involves the detection of a temperature sensitive line, normally the oxygen line λ4363. However this line is very weak and for spiral galaxies not normally visible unless the abundance is low.

2.1.2. *Nebular methods*

They involve the calculation of a grid of models to match the line intensities. They are time consuming and it quickly becomes rather unpractical to run models for every single H II region if one observed several in many galaxies.

2.1.3. *"Bright line" methods*

These are the most popular and are based in the correlation between the oxygen abundance and various oxygen and nitrogen line ratios (see Pagel et al. 1979; Edmunds and Pagel 1984; Dopita and Evans 1986). The most popular (and reliable?) is the inverse correlation between the oxygen abundance and the line ratio R_{23} defined as:

$$R_{23} = ([\mathrm{OII}]\ \lambda\lambda 3727{,}3729 + [\mathrm{OIII}]\ \lambda\lambda 4959{,}5007)\ /\ \mathrm{H}\beta,$$

as shown in Figure 1.

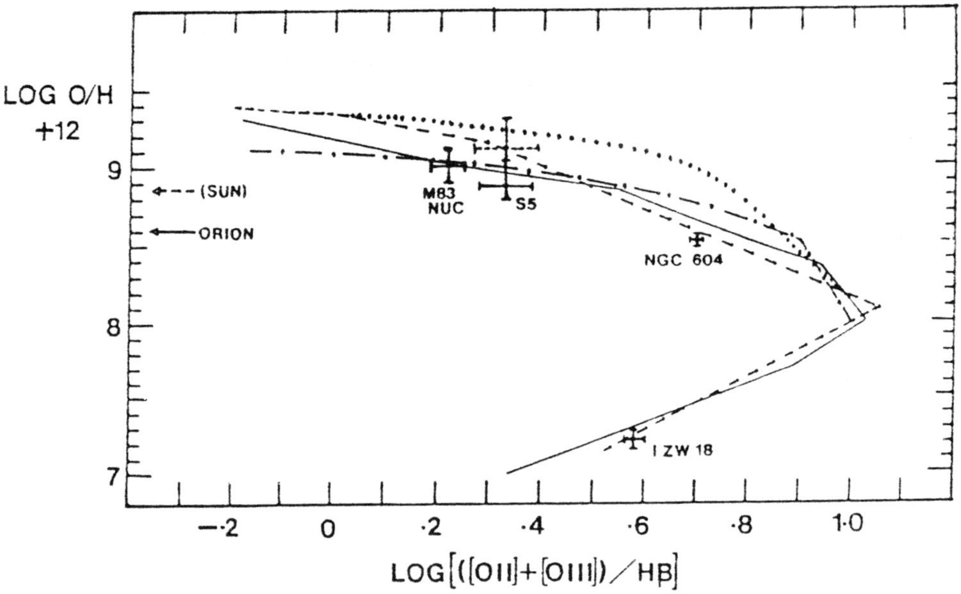

Figure 1. Relation between oxygen emission line ratio R_{23} and the oxygen abundance of a nebula as shown in Edmunds (1989). The dashed line is Edmunds and Pagel (1984), the dotted line is McCall, Rybski and Shields (1985), the dash-dot line is Dopita and Evans (1986) and the solid line Vilchez (1988). The observational points plotted are S5 from Shields and Searle (1978), as revised by Evans (1986); M83 nuclear region from Bonatto *et al.* (1989), NGC 604 from Díaz et al. (1987) and I Zw 18 from Davidson and Kinman (1985).

As the oxygen abundance decreases the cooling is less efficient so the oxygen lines appear brighter. At very low abundances where other elements take over as main coolants, there is a direct correlation so lower oxygen abundance corresponds to lower intensity lines. The duplicity of values towards the low abundance end ($(O/H) = 10^{-4}$) can be solved as the [OIII] $\lambda4363$ line should be visible. A different problem is that no reliable calibration of the curve exists at the higher abundance end (solar or higher), because here $\lambda4363$ has been detected in only a few H II regions. There are, unfortunately, a lot of spiral galaxies with H II regions which have abundances in this range. The calibration relies on the results of the nebular models used, some of which are indicated in figure 1. Recently Kennicutt and Garnett (1996) re-evaluated the metallicity gradient in M101 and found that the results are indeed sensitive to the exact calibration used for the R_{23} method.

2.2. SECOND PARAMETER

Since Aller (1942) discovered the existence of the oxygen gradients they were assumed to indicate a variation in the metallicity along the spiral disk. However it was soon realized that there was a second parameter involved, as shown by the fact that the O^{++}/O^{+} increases as the O/H decreases. This is interpreted as due to a metallicity (Z) dependent IMF by some authors (Shields and Tinsley 1976; Vilchez and Pagel 1988; Campbell 1988), which have defined a hardness parameter (η) that estimates the temperature of the ionizing stars (Vilchez and Pagel 1988). For others it indicates a variation in geometry and filling factors across the nebula (Mathis 1985; Dopita and Evans 1986).

2.3. DETERMINATION OF GRADIENTS: NORMALIZING SCALES

Once an oxygen abundance has been determined for the various H II regions across a galaxy, probably through the direct line method, and considering the various points outlined before, a negative gradient with higher abundance towards the centre is found. This is shown in Figure 2 for early and late type galaxies. The choice of the correct scale length for a spiral galaxy disk is not obvious so various normalizing scales can be used. These include the de Vaucouleurs radius, R_{25} (shown in the figure), the effective aperture R_{a}, the effective radius of the disk R_{d}, the effective radius of the spheroid R_{sp}, the radius for corotation R_{c} or the linear scale kpc. None of them reduces the observed scatter between galaxies appreciably. The origin of this scatter in abundances in a galaxy at a given radius is not clear. Undoubtedly some of it is spurious due to the inaccuracies already mentioned in the abundance determination method and observational errors. Some of it might be intrinsic. Roy and Kunth (1994) present an scenario to explain these spatial fluctua- tions, considered to be larger than expected given the apparent high efficien- cy of mixing in the ISM of most galaxies. It involves retention of enriched ejecta in evolving structures and a possible infall of low metal abundance clouds. The largest fluctuations are expected in low mass galaxies, as seen for example in I Zw 18.

In order to look for possible correlations the observed gradients are fitted with a line of the form $12 + \log \mathrm{O/H} = a + br$, where b is the observed slope and a is the projected central abundance. However as this could be a spurious value, some authors prefer to use the abundance at a certain point in the galaxy: e.g. 3 kpc or $0.4R_{25}$ or $0.8R_{\mathrm{d}}$ (ZKU94).

There is some evidence of a possible steepening of the gradients towards the inner regions of galaxies. This could be unreal due to the higher error in the determination of R_{23} at high abundance and not present when photoionization models are used to determine the oxygen abundance of

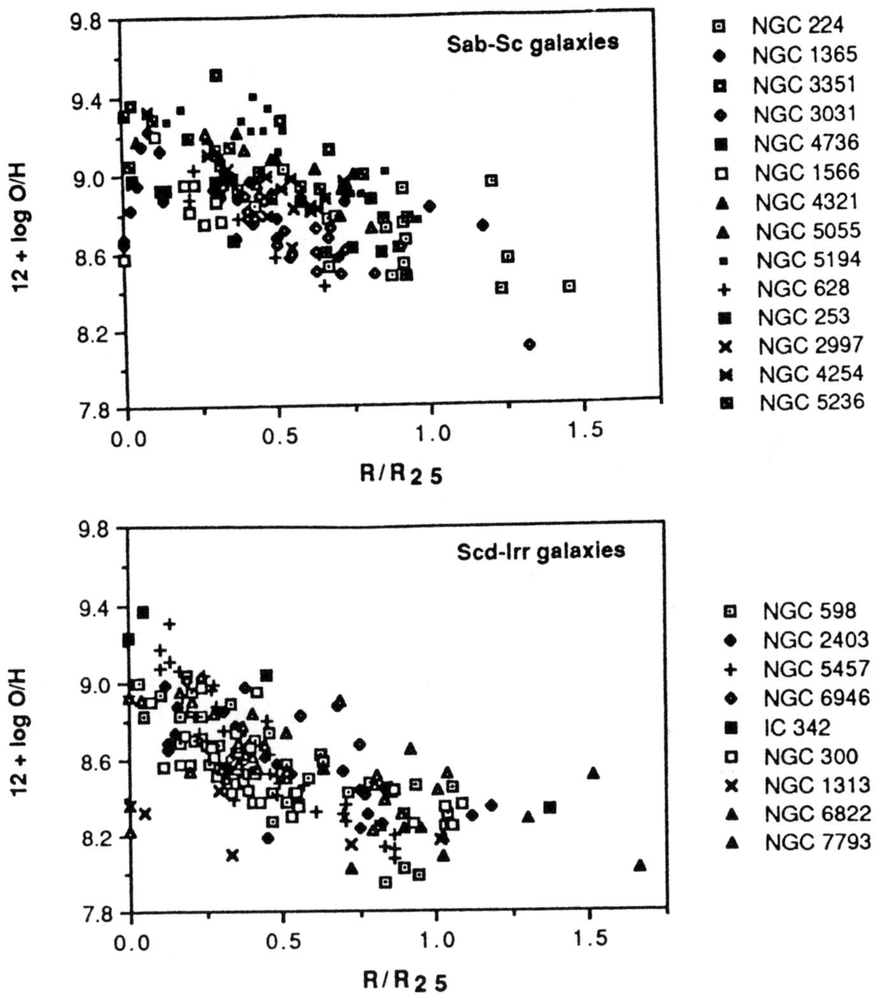

Figure 2. Oxygen abundance gradients found for early (Sab-Sc) and late (Scd-Irr) type galaxies, with the radius normalized to R_{25} . (VE92)

each H II region, as suggested by Henry and Howard (1995). Or it could be a real effect explained by a higher SFR towards the center of galaxies which consumes the inflowing gas and produces massive stars, increasing the metallicity of the ISM and steepening the gradient in the centre of a galaxy.

3. Results

3.1. BARRED VERSUS NON-BARRED GALAXIES

Barred galaxies have smaller gradients than non-barred ones (Figure 3(a)) as suggested in Vila-Costas and Edmunds (1992) and shown in the work of Martin and Roy (1994, 1995). This can be explained by a more efficient mixing of the interstellar gas across the whole galaxy disk and it is consistent with models of radial flows. The length of the bar and its ellipticity affect the gradient, with shallower gradients as these quantities increase (Martin and Roy 1994), as shown in Figure 3(b).

3.2. EARLY TYPE VERSUS LATE TYPE GALAXIES

Early type galaxies (Sa-Sb) have higher metal abundances than late types (Sc-Irr), as seen in Figure 4 (OK93). There is also some evidence in Oey and Kennicutt's data to indicate that the gradients of early type galaxies are shallower, but more work is necessary to confirm this.

The low luminosities of the H II regions in the early type galaxies, which causes them to be so difficult to detect, are due to intrinsic differences in the properties of the gas and ionizing OB associations rather than to differences in extinction as this is similar in both samples. Oey and Kennicutt (1993) have also computed photoionization models for metal- rich H II regions (0.5 - 3 $Z_{\odot}$) in order to evaluate the reliability of R_{23} in this high abundance regime. They showed its validity but also how significant uncertainties are introduced by variations in other parameters, in particular the nebular density and stellar effective temperature. Incidentally these early type galaxies do not seem to fit the mass-metallicity relation as discussed in Section 3.4.

3.3. "LOCAL" PROPERTIES OF GALAXIES

We refer as "local" properties parameters such as the surface density and brightness of disks and spheroids. Using the gradient and the central abundance, defined by the fitting to the oxygen abundances mentioned in Section 2.3, it is found that:

The gradient is correlated with the surface density (Edmunds and Pagel 1984). The correlation is better when the total (disk and spheroid) surface density is used (VE92), as shown in Figure 5.

The gradient is correlated with the surface brightness, in particular that of the spheroid (VE92), as shown in Figure 6.

The central abundance is correlated with the disk central surface density (VE92). However this quantity is itself correlated with a 'global' property: the morphological type T. (Figure 7(a)- (7c))

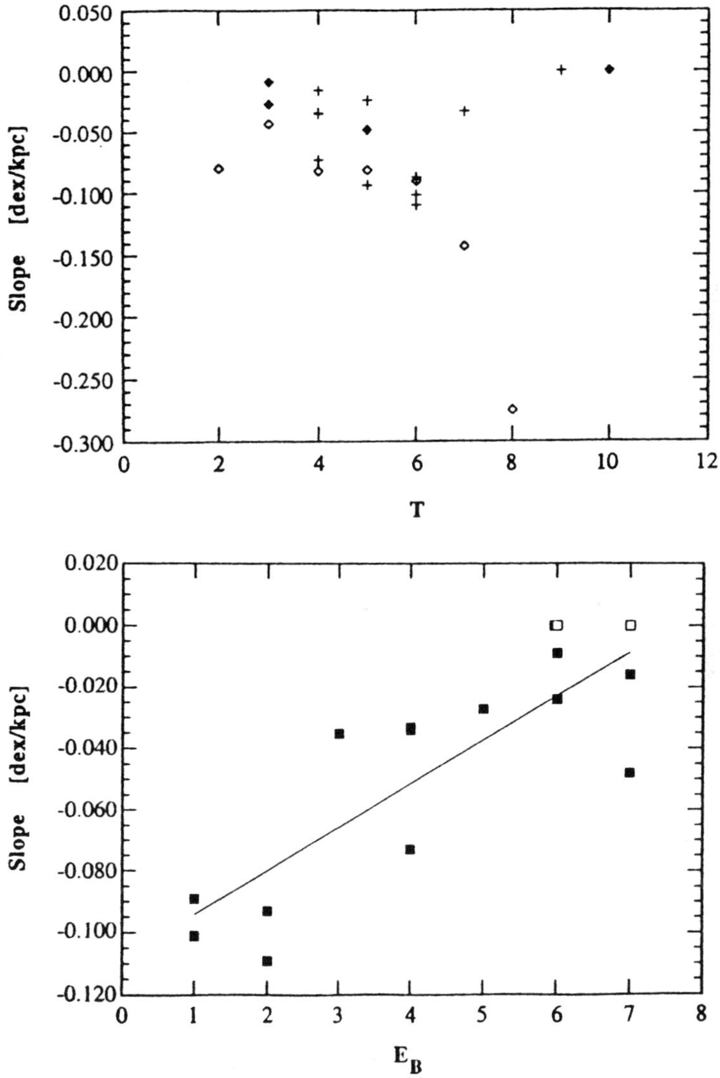

Figure 3. (a). The gradient fitted to the oxygen abundance is plotted versus morphological type. Non-barred, barred and mixed galaxies are indicated by open triangles, crosses and filled triangles respectively. 3(b). The slope of the abundance gradient is plotted versus bar ellipticity (EB). Bar strength increases with EB. (Martin and Roy 1994, 1995).

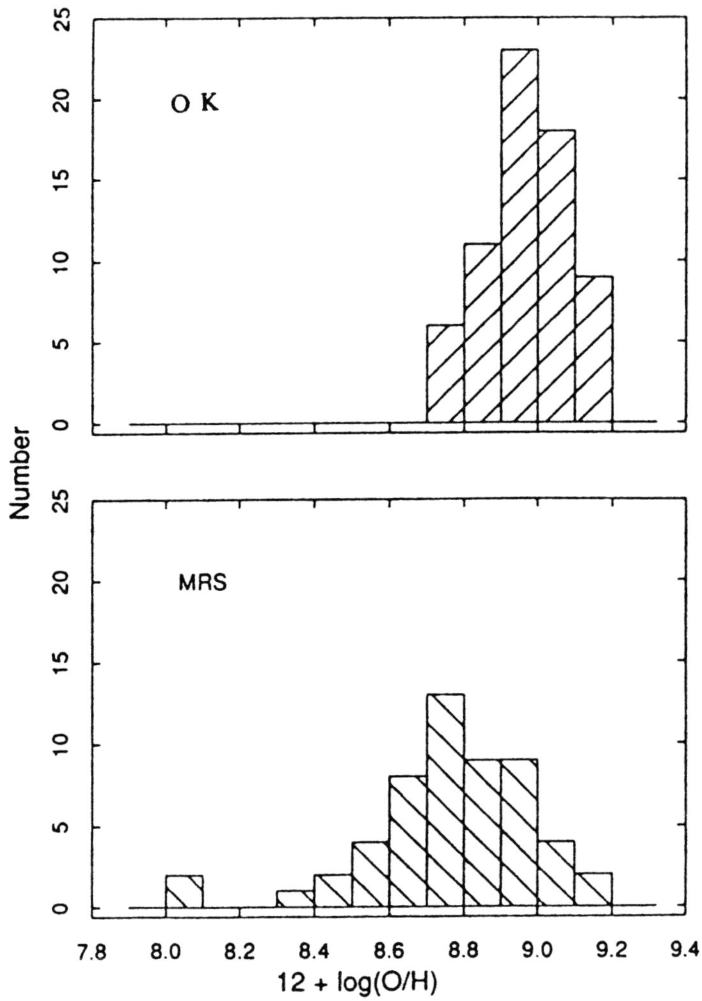

Figure 4. Comparison of abundances for Sa-Sb galaxies (top) versus types Sc and later (bottom). OK refers to Oey and Kennicutt (1993) and MRS to McCall, Rybsky and Shields (1985). (OK93).

The central abundance is NOT correlated with the disk central surface brightness (VE92), which has an approximately constant value of 21.5 for all galaxies.

Another important local property to be considered is the gas fraction along the disk of spirals. There are various uncertainties in its determination,

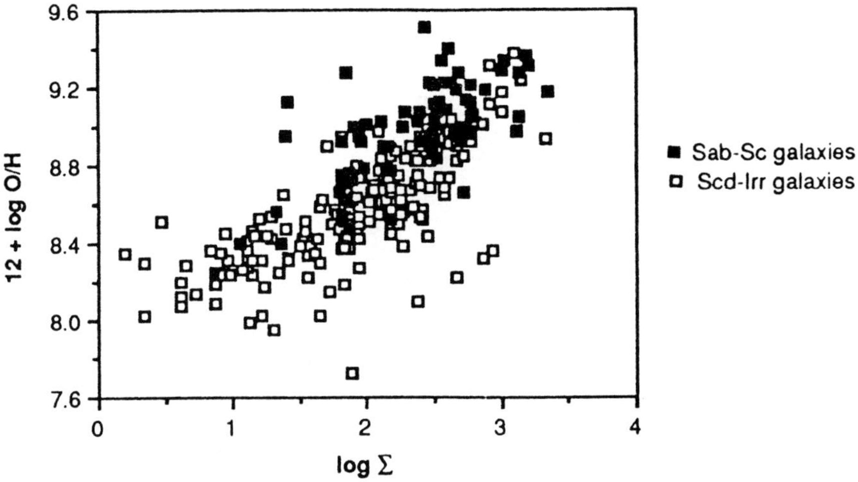

Figure 5. Correlation between the total (spheroid and disk) surface density and the oxygen abundance. Different symbols indicate early (filled squares) and late (open squares) type galaxies. (VE92)

in particular that of the CO–H_2 conversion factor , so I will discuss this on a separate section (Section 3.5).

3.4. "GLOBAL" PROPERTIES OF GALAXIES

These include the morphological type (T) and absolute magnitude (M_B) of a galaxy. The gradient of non-barred galaxies, defined in dex kpc^{-1}, is correlated with the morphological type and with the absolute magnitude, as shown in Figure 3(a). However this indicates just a disk- size dependence effect which is correlated with the morphological type, as no correlation is found if the gradient is expressed in dimensionless quantities. More massive galaxies (early T) have larger radii and then show shallower gradients than late type galaxies.

The central abundance - or the 'mean' abundance as defined for example by Zaritsky, Kennicutt and Huchra (1994) - is correlated with morphological type (Figure 7(a)) and with the absolute magnitude (Figure 8(a)). These two quantities - morphological type and absolute magnitude, expressed as either M_B or the rotational velocity (V_c) - are also correlated (Figure 8(b)).

The correlation between metallicity and absolute magnitude is present in ellipticals and dwarf spheroidals as well as irregular and spiral galaxies,

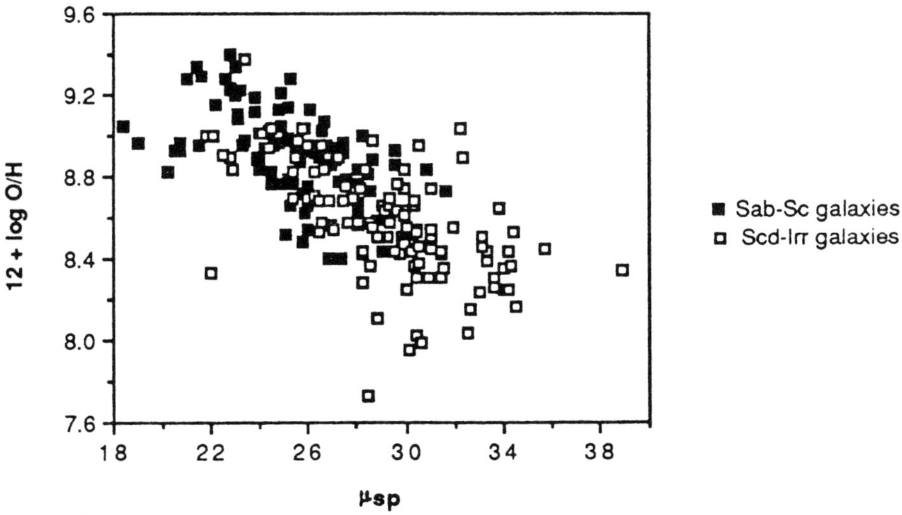

Figure 6. Correlation between the spheroid surface brightness and the oxygen abundance. Similar correlations are found for the disk and total surface brightness, although the scatter is slightly larger. Filled and open squares indicate early (Sa-Sc) and late (Scd-Irr) type galaxies. (VE92)

which shows its intrinsic importance. The same line with a displacement can be fitted to both sets of data as shown in Figure 9 (ZKH94). The abundances for E and dSph galaxies refer to the mean stellar abundances ([Fe/H]) in evolved populations rather than present-day gas phase abundances ([O/H]). The physical cause of this luminosity (mass) – metallicity correlation is unclear, in particular for large galaxies. The metallicity – morphological type correlation is more difficult to explain as T itself is not a physical parameter. Perhaps a physical parameter or evolutionary process that correl- ates with T (if not the mass) is responsible.

A discussion of other global properties like the arm class and outer gradient of the rotation curve can be found in Vila-Costas and Edmunds (1992).

3.5. GAS FRACTIONS: YIELDS

3.5.1. *Ingredients*

To carry out the study of possible correlations between gas fractions and the central intersect or the slope of the abundance gradients we need:

- The abundance gradients as we have already defined them

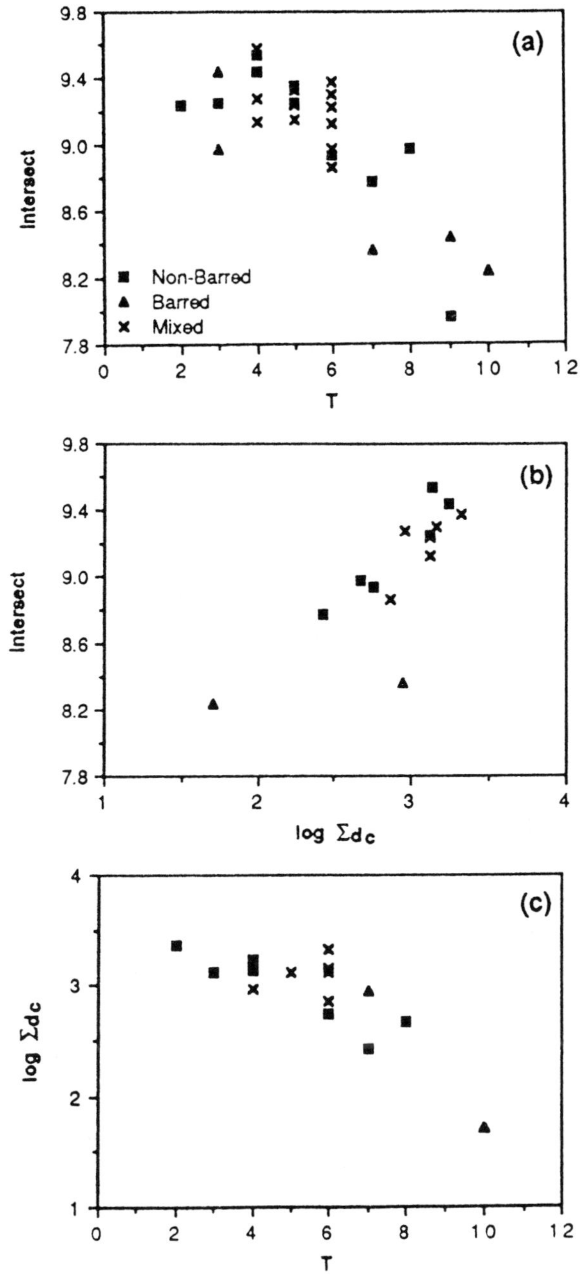

Figure 7. Correlation found between the morphological type (T) and the central abundance, extrapolated from the slopes fitted to the oxygen gradient (fig 7a) and between this central abundance and the disk central surface density (fig 7b). These two parameters (morphological type and disk central surface density) are also correlated (fig 7c). (VE92)

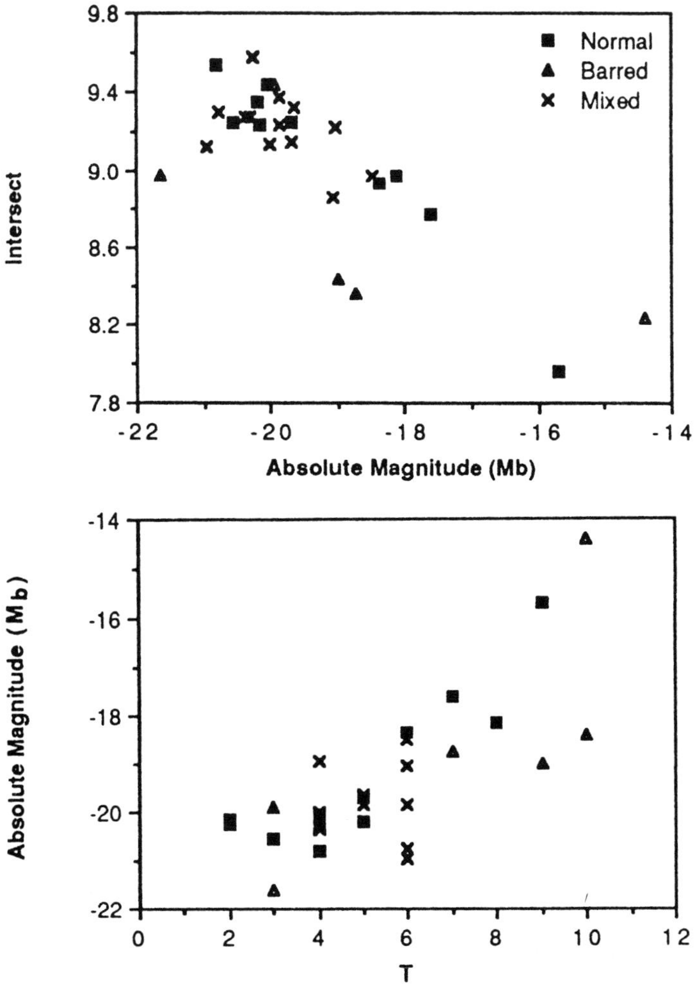

Figure 8. Correlation found between the intersect and the absolute magnitude and between the absolute magnitude and morphological type. Different symbols indicate non-barred (filled squares), barred (open triangles) and mixed (crosses) galaxies. (VE92)

- A decomposition of the rotation curves to give the disk and spheroid components of a galaxy
- The H I data
- The H_2 data.

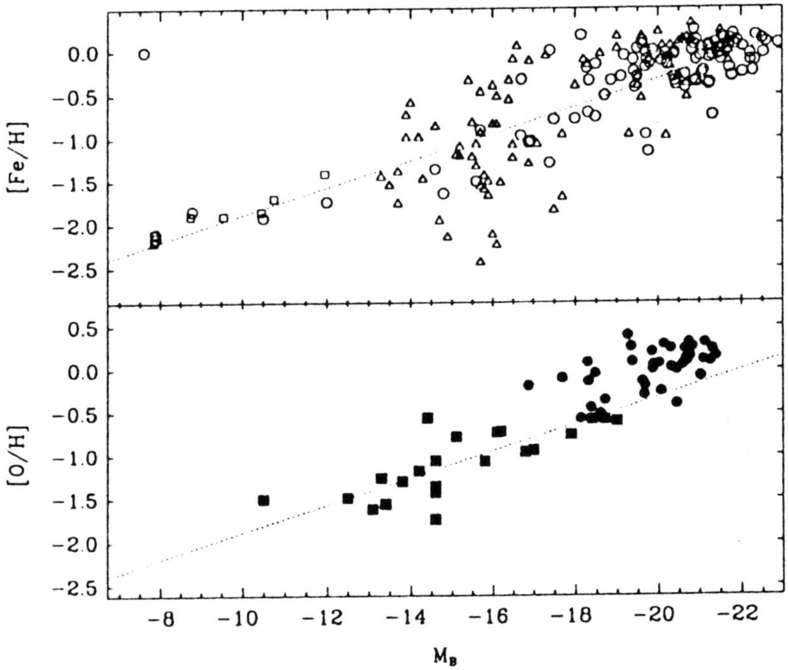

Figure 9. Luminosity and metallicity data for Local Group dEs (open squares), irregulars (filled squares), giant spiral galaxies (filled circles), and ellipticals (open circles and open triangles). References and details are given in Zaritsky, Kennicutt and Huchra (1994). The line in the upper panel is the fit provided by Brodie and Huchra (1991) for E in their sample and it is plotted on the lower panel for reference.

It is this last parameter that causes the greatest uncertainty as it has to be derived from the CO data with the use of a CO–H_2 conversion factor X (Maloney 1990, see also chapter 2, p. 36). There is no current agreement on the value of this factor and how it changes within a galaxy and between galaxies. The suggestions range from 0.8 to 4.8 $\times 10^{20}$(cm^{-2}/K km s^{-1}) with a possible dependence on metallicity ($1/Z$, $1/Z^2$)

We have the required data for 8 galaxies (VE92) and carried out an attempt to see if a particular conversion factor gave a "better", more 'smooth' profile for the total gas content in a galaxy. For example see Figure 10 for the galaxy NGC 598. Perhaps not surprisingly no clear answer emerged. We decided to use for each galaxy the conversion factor that gave the smoothest exponential total profile for H I + H_2.

There is a correlation between the oxygen abundance and the H_2 gas fraction (H_2/Gas) (Tosi and Díaz 1991; Vila-Costas and Edmunds 1992). The physical significance of this is not clear and it might have more to do with the CO–H_2 conversion factor than with star formation and element synthesis.

3.5.2. *Yield*

This is defined as a quantity which represents the amount of newly synthesized heavy elements returned to the interstellar medium per amount of mass locked up in long lived stars, in each generation of star formation. In the instantaneous recycling approximation it is related to the metallicity Z and to the gas fraction f. To investigate the possibility of a varying yield an effective yield is defined as the yield that would be deduced if a region of a galaxy were assumed to be behaving as the simple chemical evolution model (Edmunds 1990),

$$p_{\rm eff} = Z \;/\; \ln\,(1/f)$$

where $\log Z = 1.42 + \log$ O/H and f = Total gas mass / total mass stars and gas > 0.01

The results obtained from Figure 11 show an effective yield within a small range, $0.004 < p_{\rm eff} < 0.012$, decreasing with radius and/or increasing with metallicity. The extent of this variation is rather dependent on the adopted CO–H_2 conversion factor. A more detailed account can be found in Vila-Costas and Edmunds (1992).

4. Conclusions

The study of the abundances in spiral galaxies has advanced a lot in the last several years. As more/better data are made available we hope to get more correlations to define the important parameters that determine the observed oxygen gradients.

The main results so far indicate how barred galaxies have shallower gradients than non-barred ones through a more efficient mixing of the interstellar gas. This gradient is related to the length and ellipticity of the bar. Early type galaxies show smaller gradients than late type ones. The study of early type galaxies is complicated as their H II regions are intrinsically less luminous - this is not an extinction effect.

The gradient found in spiral galaxies is correlated with their surface density and surface brightness. There is a correlation between the central abundance, as defined through the fitting of the observed oxygen gradients, and the central surface disk density, which is in itself correlated with the morphological type. No correlation is found with the central surface bright-

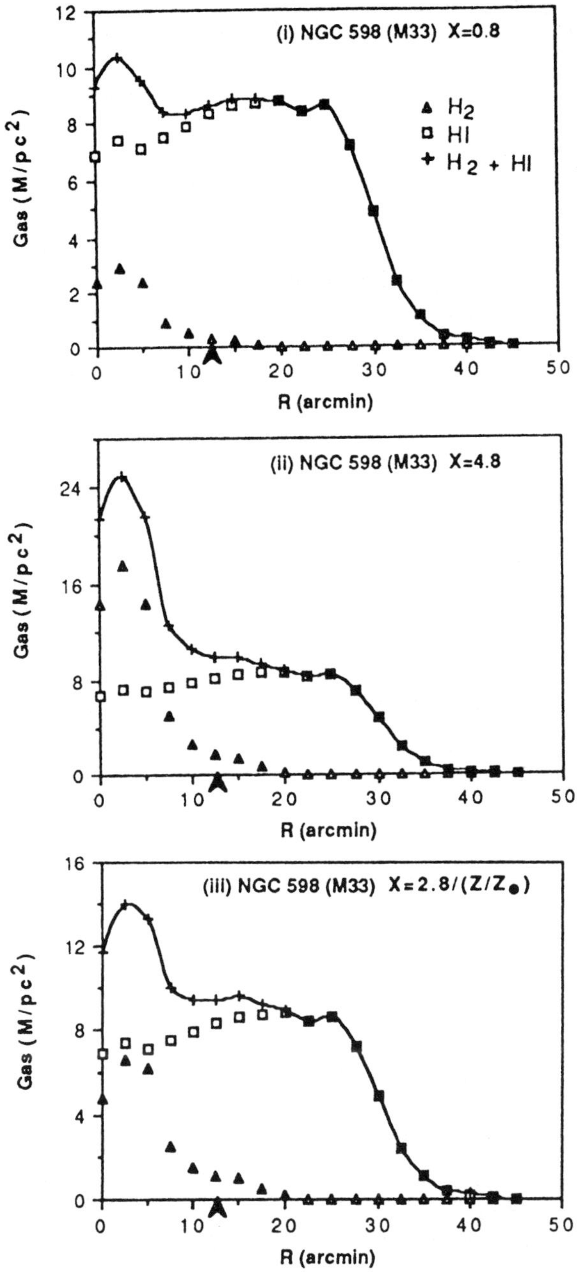

Figure 10. Example of the gas distribution for the galaxy NGC 598. HI , H_2 and total (HI +H_2) distributions are indicated by the open squares, open triangles and filled diamonds respectively. Various CO–H_2 conversion factors have been used to obtain H_2 from the CO distribution. These are: (i) 0.8 (ii) 4.8 (iii) 2.8/Z. The value for $Z_\odot = 0.013$ has been used. The radius from which $Z < Z_\odot$ is indicated. (VE92)

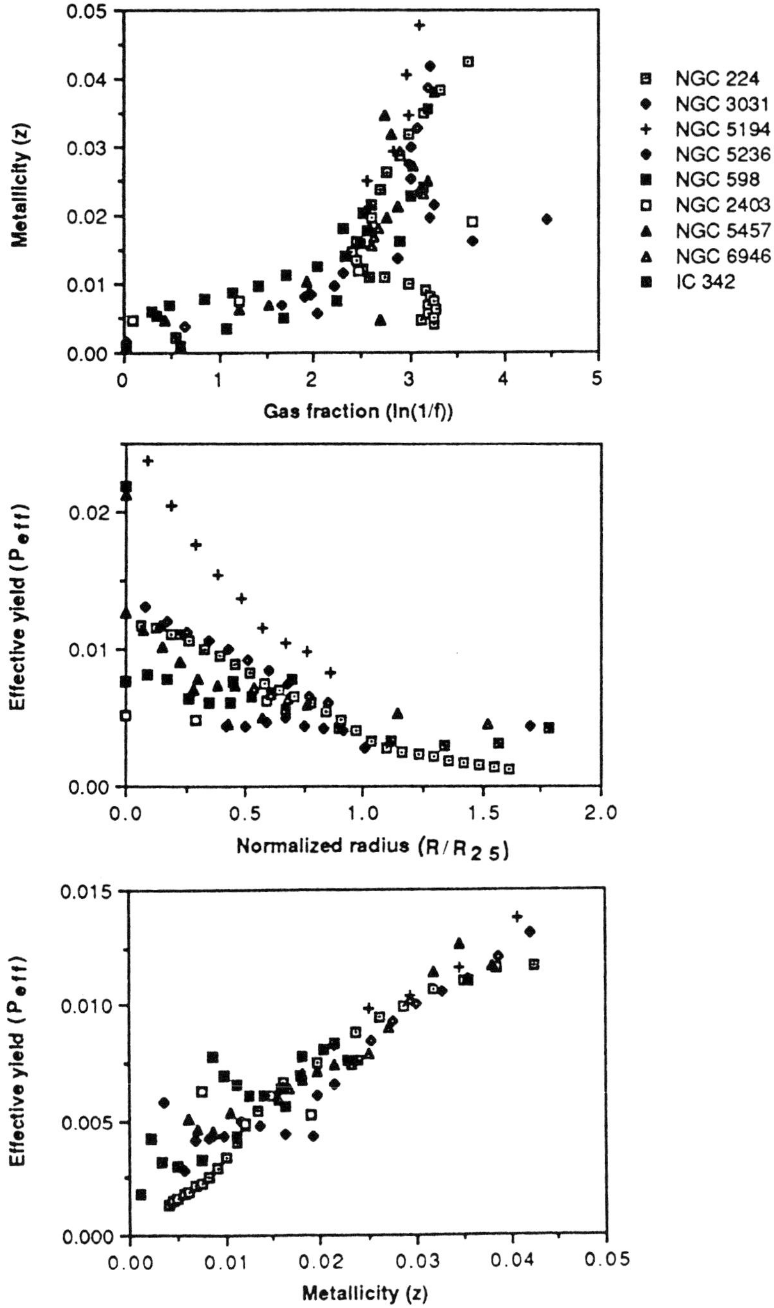

Figure 11. Distribution of (a) Z vs. ln (1/f) (b) Effective yield, $p_{\rm eff} vs. R/R_{25}$ (c) Z vs. $p_{\rm eff}$, using the "best" CO–H_2 conversion factor (VE92). Those points where the gas fraction was less than 1%, i.e. ln (1/f) > 4.5 have been suppressed.

ness which has an almost constant value of 21.5 for all galaxies studied here.

When the gradient is expressed in dex/kpc there is a correlation with the morphological type and with the absolute magnitude of the parent galaxy. However this is just a disk-size dependence on morphological type, with more massive galaxies (early T) having larger radii and so shallower gradients and late type galaxies of smaller sizes showing steeper gradients.

The central abundance – or the "mean" abundance – is correlated with the morphological type and the absolute magnitude, which are also correlated with each other. The correlation with the absolute magnitude is present in ellipticals, dEs and irregulars as well as spiral galaxies.

To look for correlations of the oxygen abundance with the gas fraction in galaxies we need a CO–H_2 conversion factor to obtain the H_2 distribution from the CO data. Hopefully in the near future this point will be solved. With the available data at present the effective yield seems to cover a range between 0.004 and 0.012, and to decrease with radius and/or increase with metallicity along the disk of a galaxy.

References

Aller, L. H. 1942, *Ap.J.* , 95, 52.

Bonatto, C., Bica, E. and Alloin, D. 1989, *Ast.Ap.* , 226, 23.

Brodie, J. and Huchra, J. 1991, *Ap.J.* , 379, 157.

Campbell, A. 1988, *Ap.J.* , 335, 644.

Clarke, C. J. 1989, *M.N.R.A.S.* , 238, 283.

Davidson, K. and Kinman , T. D. 1985, *Ap.J.Suppl.* , 58, 321.

Díaz, A. I. 1989, in *Evolutionary Phenomena in Galaxies*, eds. J. E. Beckman and B. E. J. Pagel, Cambridge University Press, p. 377.

Díaz, A. I. and Tosi, M. 1984, *M.N.R.A.S.* , 208, 365.

Díaz, A. I., Terlevich, E., Pagel, B. E. J., Vilchez, J. M. and Edmunds, M. G. 1987, *M.N.R.A.S.* , 226, 19.

Dinerstein, H. L. 1990, in *The Interstellar Medium in Galaxies*, eds. H. A. Thronson Jr. and J. M. Shull, p. 257

Dopita, M. A. and Evans, I. N. 1986, *Ap.J.* , 307, 431.

Edmunds, M. G. 1989, in *Evolutionary Phenomena in Galaxies*, eds. J.E. Beckman and B.E.J. Pagel, Cambridge University Press, p. 356

Edmunds, M. G. 1990, *M.N.R.A.S.* , 246, 678.

Edmunds, M. G. and Pagel, B. E. J. 1984, *M.N.R.A.S.* , 211, 507.

Edmunds, M. G. and Phillipps, S. 1989, *M.N.R.A.S.* , 241, 9.

Evans, I. N. 1986, *Ap.J.* , 309, 544.

Garnett, D. R. and Shields, G. A. 1987, *Ap.J.* , 317, 82.

Güsten, R. and Mezger, P. G. 1982, *Vistas in Astr.*, 26, 159.

Henry, R. B. C. and Howard, J. W. 1995, *Ap.J.* , 438, 170.

Kennicutt, R. C., Oey, M. S., Zaritsky, D. and Huchra, J. P. 1993, *Revista Mexicana de Astr. y Astrofisica* 27, 21.

Kennicutt, R. C., and Garnett, D. R. 1996, *Ap.J.* , 456, 504.

Lacey, C. G. and Fall, S. M. 1985, *Ap.J.* , 290, 154.

Maloney, P. 1990, in *The Interstellar Medium in Galaxies*, eds. H. A. Jr Thronson and J. M. Shull, Kluwer Academic Publishers, p. 493.

Martin, M. and Roy, J.R. 1994, *Ap.J.* , 424, 599.
Martin, M. and Roy, J.R. 1995, *Ap.J.* , 445, 161.
Mathis, J. S. 1985, *Ap.J.* , 291, 247.
Mayor, M. and Vigroux, L. 1981, *Ast.Ap.* , 98, 1.
McCall, M. L. 1982, Ph. D. Thesis, Univ. of Texas, Austin.
McCall, M. L., Rybski, P. M. and Shields, G. A. 1985, *Ap.J.Suppl.* , 57, 1.
Oey , M. S. and Kennicutt, R. C. 1993, *Ap.J.* , 411, 137. (OK93)
Pitts, E. and Tayler, R. J. 1989, *M.N.R.A.S.* , 240, 373.
Pagel, B. E. J. and Edmunds, M. G. 1981, *Ann.Rev.Ast.Ap.* , 19, 77.
Pagel , B. E. J., Edmunds, M. G., Blackwell, D. A., Chun, M. S. and Smith, G. 1979, *M.N.R.A.S.* , 189, 95.
Phillipps, S., Edmunds, M. G. and Davies, J. 1990, *M.N.R.A.S.* , 244, 168.
Roy, J-R. and Kunth, D. 1995, *Ast.Ap.* , 294, 432
Shields, G. A. and Searle, L. 1978, *Ap.J.* , 222, 821.
Shields, G. A. and Tinsley, B. M. 1976, *Ap.J.* , 203, 66.
Shields, G. A., Skillman, E. D. and Kennicutt, R. C. 1991, *Ap.J.* , 371, 82.
Sommer-Larsen, J. and Yoshii , Y. 1989, *M.N.R.A.S.* , 238, 133.
Vila-Costas, M. B. and Edmunds, M. G. 1992, *M.N.R.A.S.* , 259, 121. (VE92)
Vila-Costas, M. B. and Edmunds, M. G. 1993, *M.N.R.A.S.* , 265, 199.
Vilchez, J. M. 1988, Ph.D. Thesis, IAC, La Laguna, Spain.
Vilchez, J. M. and Pagel, B. E. J. 1988, *M.N.R.A.S.* , 231, 257.
Zaritsky, D., Kennicutt, R.C. and Huchra, J. P. 1994, *Ap.J.* , 420, 87. (ZKU94)

FROM GAS TO STARS: REGULATION OF STAR FORMATION

R. C. KENNICUTT
Steward Observatory
University of Arizona
Tucson, AZ,
U.S.A.

1. Introduction

Understanding the physical processes that govern the conversion of interstellar gas into stars is crucial for a host of fundamental problems: galactic structure, the nature of the Hubble sequence, galaxy formation and evolution, and chemical evolution. Although our understanding of large-scale star formation and its relationship to the interstellar medium (ISM) is at a primitive state, a recent flurry of observations and theoretical modelling has provided important new insights into the star formation law and its physical basis.

The roots of this field trace to the seminal paper of Schmidt (1959), which introduced the star formation rate (SFR) vs. gas density power law that is still used today. The Schmidt law is an adequate description for some aspects of the problem, but observations of nearby galaxies have revealed environments where a single power-law parametrization clearly breaks down. A significant breakthrough came with the observation of star formation thresholds, which when combined with a Schmidt law at high densities reproduces the wide range of star formation properties of galaxies. The tentative association of these thresholds with large-scale gravitational instabilities in the ISM provides a physical framework for the observed star formation law, and the beginnings of a quantitative model for global star formation in galaxies. Several important elements of this picture— the physical nature of the star formation thresholds, the physical regulation of the SFR above the threshold, and the role of feeback between stars and gas in the star formation law— remain poorly understood and are the focus of much current research.

J. M. van der Hulst (ed.), The Interstellar Medium in Galaxies, 171–195.

In this paper I review recent progress in our understanding of the empirical form and physical basis of the star formation law, and identify some of the key issues for future work. Section 2 reviews the successes and failures of the Schmidt law, with emphasis on those problems for which the Schmidt law *does* serve as a useful parametrization of the SFR. Section 3 summarizes recent work on star formation thresholds: observational evidence, the gravitational stability model, and the successes and failures of that model. The physical nature of these "star formation" thresholds is discussed in Sec. 3.4. While thresholds may provide physical insight into why stars stars do *not* form in galaxies, they leave unanswered the question of what controls the SFR at densities above the threshold. Section 4 reviews some recent ideas on this subject, with special emphasis on self-regulating models. For background material on earlier developments in this field I refer the reader to reviews by Larson (1987, 1988) and Kennicutt (1990).

2. The Schmidt law

Thirty-five years after its introduction the Schmidt law remains as a useful parametrization of the SFR for many applications, and the basis of nearly all subsequent work. The original Schmidt (1959) law related the volume densities of young stars and gas in the Galactic disk:

$$\rho_{sf} = a\rho_{gas}^{n}$$

but for external galaxies, where projected densities are observed, it is conventional to adopt a surface density law:

$$\Sigma_{sf} = A\Sigma_{gas}^{N}$$

It is important to recognize the physical distinction between the observed and true (3D) laws. Fortunately in most environments the slope of the power law is small enough that the distinction is not crucial.

Since 1959 roughly 50 papers have been published which attempt to measure the value of N, by correlating the surface densities of a young Population I stellar tracer with gas column densities. Compilations and discussions of this work can be found in Berkhuijsen (1977), Madore (1977), Freedman (1984), and Kennicutt (1990). My own updated compilation is summarized in Figure 1, which shows the distribution of derived values of N. The values cover a wide range, with a broad peak between $N = 0.8$ and 2.5 and a full range of $N = \pm 3.5$. This large dispersion can be attributed partly to observational factors, such beam smearing effects or fitting to only the atomic or molecular gas densities. However much of the scatter is real, being caused by such factors as deviations from a power law or spatial variations in N across a galaxy. The failure of this wide body of

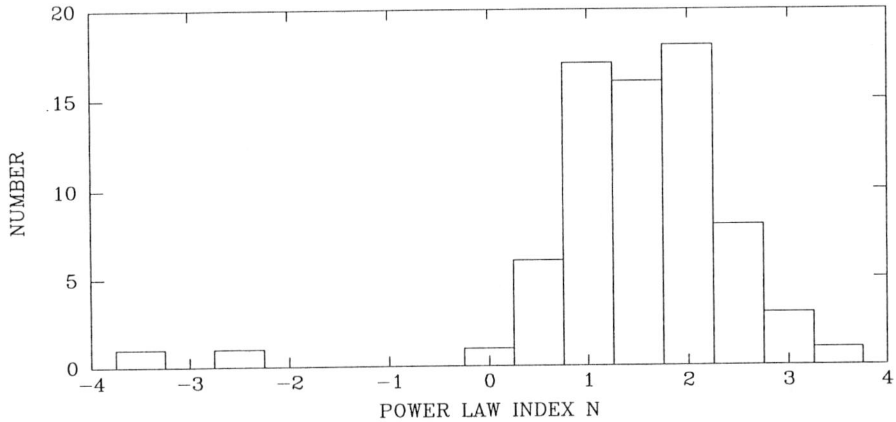

Figure 1. Best fitting values of Schmidt law index N, from a compilation of the literature.

observations to converge on a single Schmidt law is evidence in itself for the limitations of a power-law parametrization.

Nevertheless there are applications for which a Schmidt law *does* provide a useful description of the SFR, especially when applied to the integrated SFRs of galaxies. This is illustrated in Figure 2, which shows the relationship between the average Hα disk surface brightness and the average gas surface density (molecular and atomic) for a sample of nearby spirals (Kennicutt & Martin 1997). Each point represents a single galaxy, where the Hα intensities (which should scale with the SFR) and gas densities have been averaged over the optical diameter of the disk. Although there is considerable scatter in the relation, the correlation demonstrates that a Schmidt law with $N \sim 1.5$ provides a reasonable parametrization of the global evolution of disks.

Several authors have analyzed the dependence of the disk-averaged SFR on the molecular and atomic surface densities separately. For example Figure 3 shows the dependence of ultraviolet surface brightness, as measured by the FAUST experiment, on average total, H_2, and H I surface densities (Deharveng *et al.* 1994). The coupling between the SFR and molecular surface density is surprisingly weak, a trend that has been observed in several other studies (e.g., Buat *et al.* 1989, Kennicutt 1989, Buat 1992, Boselli 1994), and is probably related to variations in the CO–H_2 conversion factor (see also chapter 2, p. 36). On the other hand, the SFR is tightly correlated with H I surface density, even though H I often comprises only a small fraction of the (mostly molecular) disk within the star forming radius. This raises the question of whether the SFR is being determined by

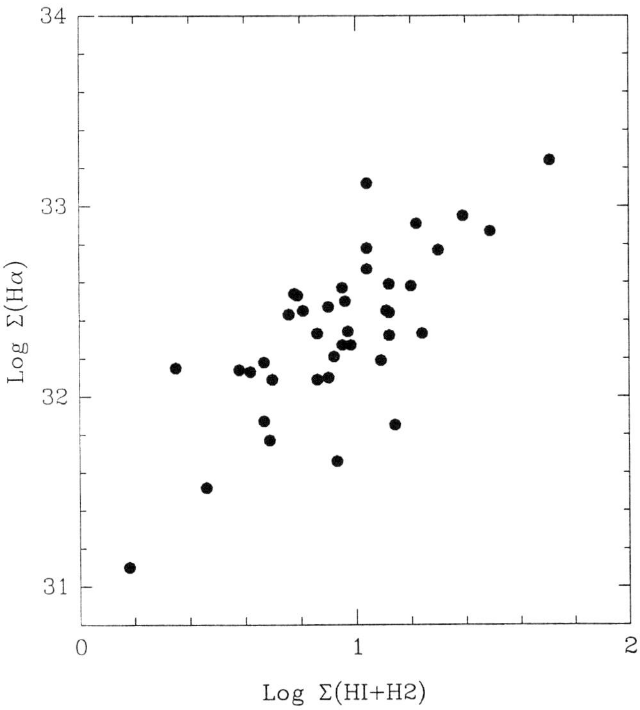

Figure 2. Relationship between disk-averaged Hα surface brightness, proportional to the SFR per unit area, and average gas surface density for nearby spiral galaxies, from Kennicutt & Martin (1997).

the gas density, or whether it is the SFR which is regulating the H I density, for example via photodissociation of molecular hydrogen (e.g., Shaya & Federman 1987; Tilanus & Allen 1989; Knapen & Beckman, 1996). This example illustrates the potential pitfalls in inferring causal relations from an observed correlation, without a full understanding of its underlying physical basis.

The Schmidt law provides a less deterministic description of the SFR on smaller physical scales. This is illustrated in Figure 4, which shows the correlation of Hα surface brightness on H I column density in cells of 220 pc in the LMC (Kennicutt *et al.* 1995). When the data are binned in column density a monotonic Schmidt law is apparent, with $N = 1.75 \pm 0.3$, consistent with previous studies (e.g., Tosa & Hamajima 1975). However the distribution of individual points is a virtual scatter diagram, with the correlation evident only when the data are binned. Compare this scatter with the relatively tight correlations in Figs. 2–3. Similar results were found in a study of M33 by Freedman (1984). These results suggest that the

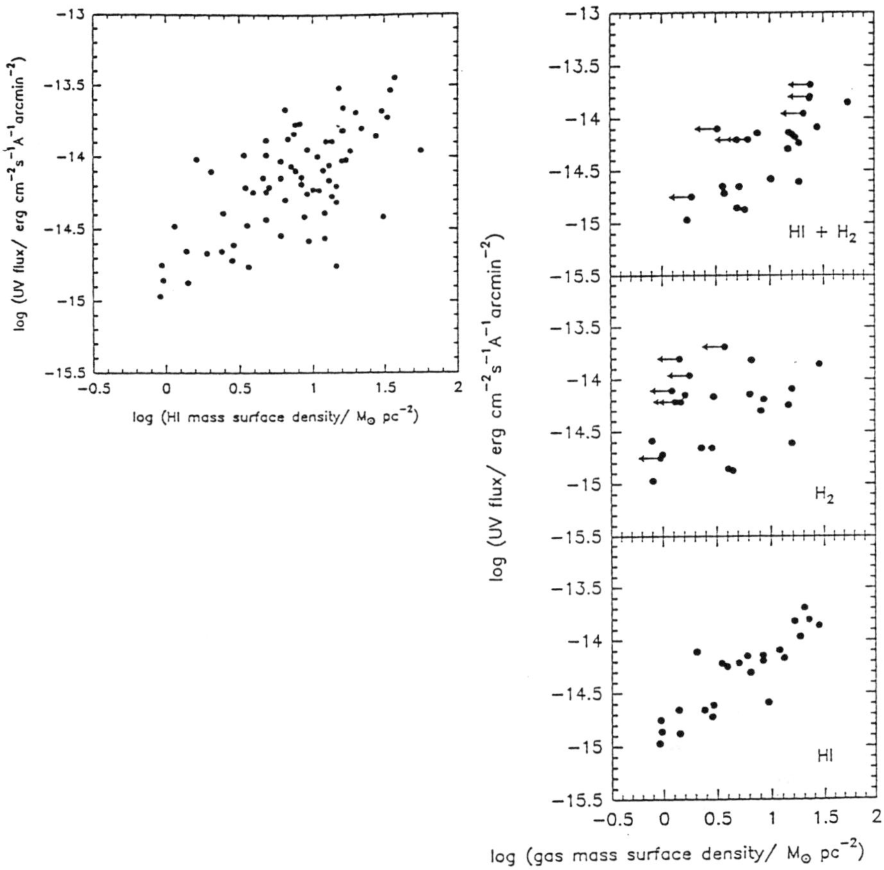

Figure 3. Correlation of mean UV surface brightness on gas surface density, showing the separate dependences on total, molecular, and atomic gas surface densities. From Deharveng *et al.* 1994.

Schmidt law should be regarded at best as a statistical description of the star–ISM relation. The change in the character of the relation with physical scale should not be surprising, because the physical processes which underlie the observed law certainly change from the parsecs scale, where internal cloud processes (e.g., star formation efficiency) dominate, to galactic scales where large scale disk instabilities and cloud formation/destruction processes are most important.

There are several physical environments where the Schmidt law, if it applies at all, changes form radically under different conditions. The global SFRs in active star forming disks, for example, are governed by a relatively mild ($N \sim$ 1–1.5) Schmidt law, as shown in Figures 2–3. The relatively

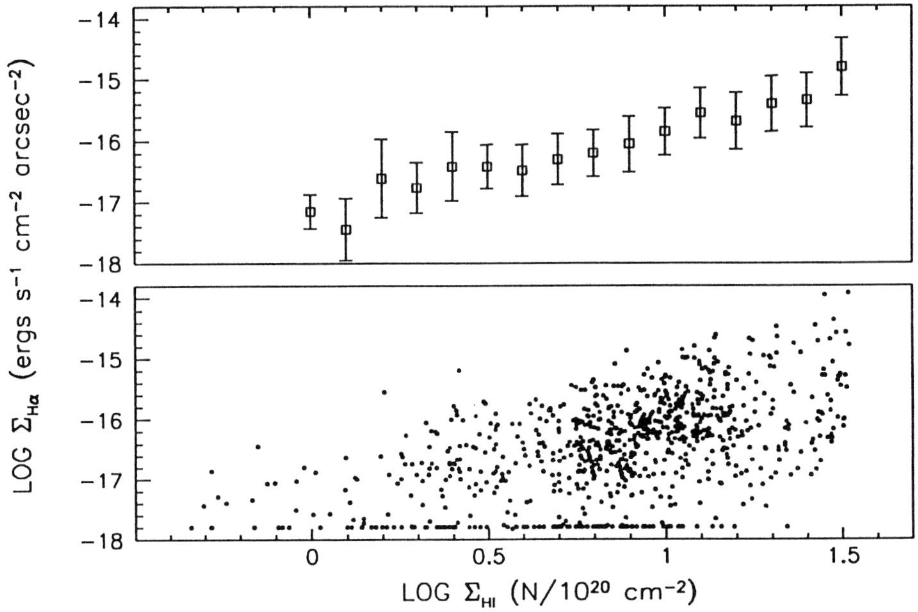

Figure 4. Hα vs. HI correlation for the LMC. Points represent individual 220 pc cells, while the bars show mean trends when the points are binned in column density. From Kennicutt *et al.* (1995).

slow evolution of the stellar birthrate in the solar neighborhood and in other disks also requires a weak dependence of SFR on gas density (e.g., Kennicutt *et al.* 1994). On the other hand, highly nonlinear ($N \gg 1$) Schmidt laws are needed to account for the orders-of-magnitude increases in the SFRs of starburst galaxies and interacting galaxies, or the strong arm-interarm constrasts in star formation observed in grand-design spirals such as M51 (e.g., Rand 1993). Similar nonlinearities are required to account for the *absence* of star formation in low surface brightness spirals or gas-rich S0 galaxies, which fail to form stars despite possessing gas masses that often are several times that of the Galaxy. This wide range of linear and nonlinear behaviors must hold an important clue to the physical processes which regulate the SFR. Fortunately a simple modification to the Schmidt law— the introduction of a Jeans-like star formation threshold— can account for this wide range of phenomena within a single quantitative framework.

3. Star formation thresholds

3.1. OBSERVATIONS

There is now extensive observational evidence that the SFR vs. gas density relation turns over abruptly below a critical threshold value (e.g., Davies

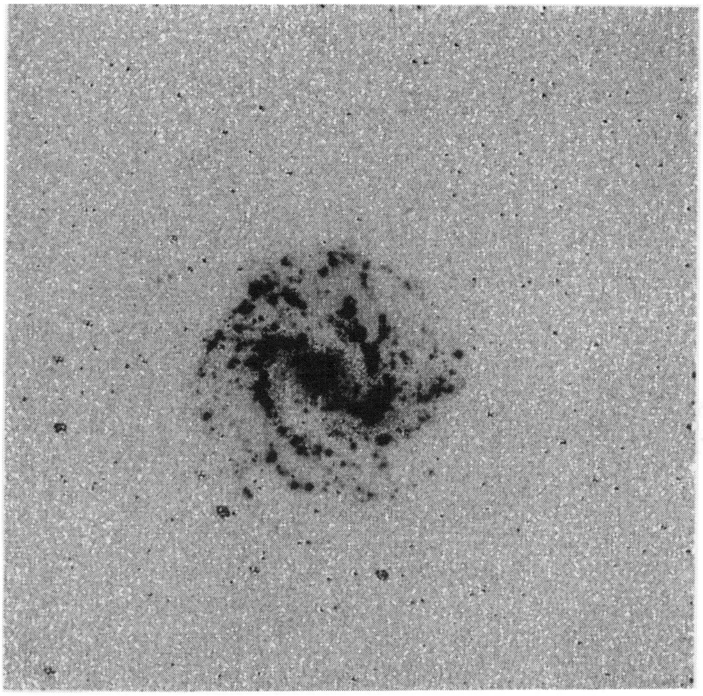

Figure 5. Deep Hα image of M83, showing sharp outer threshold to the star forming disk. The diameter of the Hα disk is $\sim$10$'$, compared to an HI diameter of 40$'$. The spots are large radius are artifacts of the foreground star subtraction.

et al. 1976, Gallagher & Hunter 1986, Guiderdoni 1987, Skillman 1987, van der Hulst *et al.* 1987, Kennicutt 1989, Taylor *et al.* 1994). A pictorial example is shown in Figure 5, which shows a high-contrast Hα image of the nearby SBc galaxy NGC 5236 (M83). There is nothing subtle about the threshold; an abrupt edge to the active star forming disk is evident, even though the H I extends to 3 times the optical radius (Rogstad *et al.* 1974). Sporadic star formation is often observed beyond the threshold radius, but at a level which is far below the extrapolation of the inner exponential disk. Often the threshold is abrupt, as in M83, but in other cases the transition region is spatially resolved. Usually the threshold occurs near the optical radius of the disk (R_{25}), but in some cases it extends to near the limit of the gas distribution (e.g., M101) and in others well inside the stellar disk (e.g., NGC 4736, NGC 4569). As discussed by Kennicutt (1989) there is no clear association of the Hα thresholds with the transition between a molecular and atomic-dominated disk, but the available observations are barely adequate to conduct such a test (see Sec. 3.4).

The effect of the thresholds on the observed star formation law is shown in Figure 6, taken from Kennicutt & Martin (1997). The top panel shows the dependence of Hα surface brightness on (total) gas surface density, as measured in radial rings for several active star-forming galaxies. Note that galactocentric radius varies over the length of each profile, so factors other than gas density may be influencing the SFR. At high densities the relations follow a very rough Schmidt law with mean $N \sim 1.3$, but the SFR drops abruptly at the threshold density. The second panel shows the same relation for a sample of early-type spirals, with detectable but low SFRs across their disks. The SFR–density relation is much less well defined in this case, but note that the qualitative behavior of the law is similar to that seen in the threshold regime in the active star forming galaxies. The thresholds occur at surface densities of order a few $M_\odot$ pc^{-2}, or $\sim 10^{21}$ H cm^{-2}, but with a significant variation between galaxies, with roughly a tenfold range over the entire sample.

3.2. THE GRAVITATIONAL STABILITY MODEL

Kennicutt (1989) suggested that the observed star formation thresholds might be caused by large-scale gravitational instabilities in the gas disks. Such a suggestion is not new. Quirk (1972), Elmegreen (1979), Cowie (1981), Larson (1983), and Lacey & Fall (1985) explored the possibility of gravitationally-induced cloud or star formation thresholds, but insufficient data were available for a rigorous test of the model. Kennicutt (1989) tested the model using the stability criterion of Safronov (1960) and Toomre (1964):

$$\Sigma_c = \frac{\kappa \sigma_{\rm v}}{\pi G}$$

$$Q \equiv \frac{\Sigma_c}{\Sigma_{gas}}$$

where Σ_c is the critical surface density, κ is the epicyclic frequency at a given radius, and $\sigma_{\rm v}$ is the velocity dispersion of the gas. This of course is an idealization of any actual threshold for instabilities in a realistic galactic disk, but it should at least approximate the actual criterion for any gravitational instability in regions where the disk rotation velocity is roughly constant (see Sec. 3.4). The observation that the gas velocity dispersion in disks is roughly isothermal (e.g., van der Kruit & Shostak 1984) allows us to apply the Toomre criterion with a minimum of free parameters. For a constant value of $\sigma_{\rm v}$ the critical density scales as v/r, where v is the circular rotation velocity (constant over most of the disk) and r is galactocentric distance. Hence Σ_c falls off roughly as $1/r$ over most

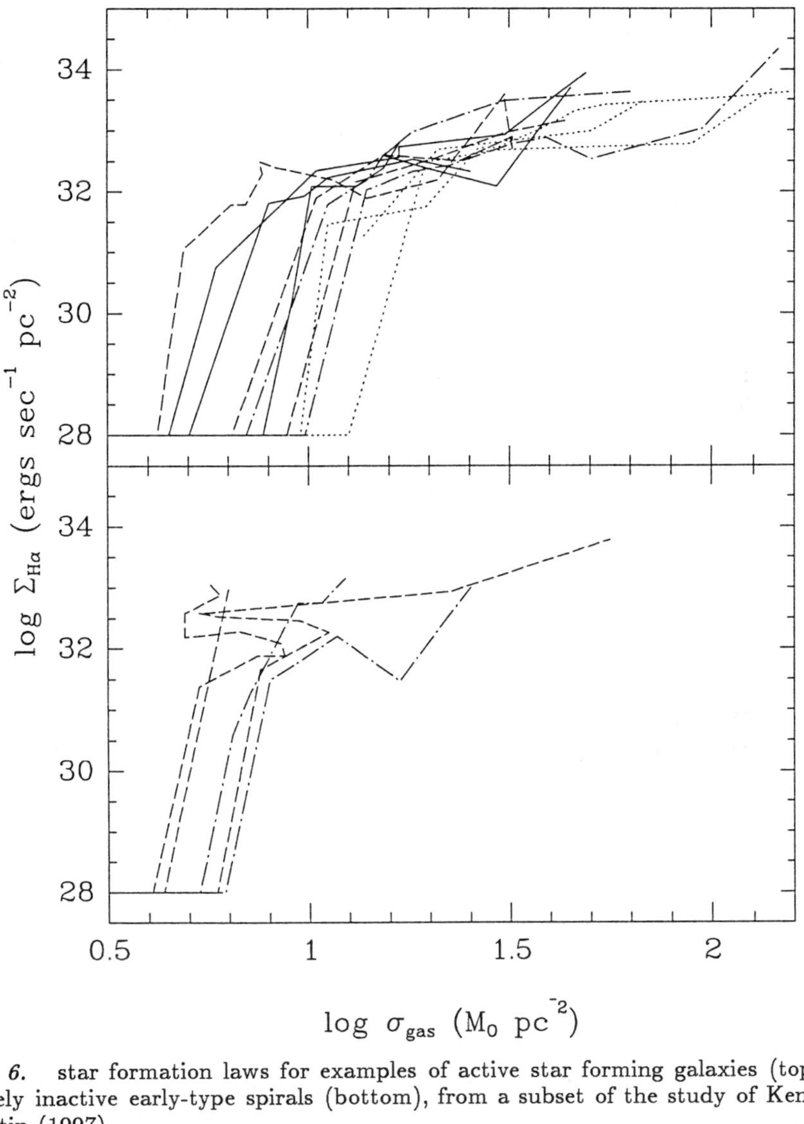

Figure 6. star formation laws for examples of active star forming galaxies (top) and relatively inactive early-type spirals (bottom), from a subset of the study of Kennicutt & Martin (1997).

of the disk, and the SFR will be dictated by the relation of this decline to the actual radial density profile of the gas in the disk.

Kennicutt (1989) computed $\Sigma_c(r)$ and $Q(r)$ for ~20 galaxies with available Hα CO, H I data and rotation curves, assuming a constant gas velocity dispersion $c = 6$ km s^{-1}. Figure 7, taken from that paper, shows the radial run of $\Sigma_{gas}/\Sigma_c = Q^{-1}$ for the star forming galaxies (top) and early-type, low-SFR galaxies (bottom) in the sample. The radial coordinates are normalized to the observed threshold radius (top) or optical radius (bottom). In the star forming galaxies (top) the Q criterion readily separates the regions of active star formation, and the curves all pass through the stability limit ($Q \sim 1.5$) near the observed threshold radii. The same Q criterion successfully explains the absense of global star formation in the early-type disks; in those few regions where $Q < 1.5$ rings of bright H II regions are observed. The deviation of the best fitting stability value from $Q = 1$ could be real or reflect an incorrect choice of the gas velocity dispersion.

The gravitational stability criterion provides a natural physical framework for interpreting the observed star formation law and the gas distributions in disks. One expects three distinct star forming regimes: 1) At high gas densities relative to the threshold ($Q \ll 1$) cloud formation proceeds without inhibition, and global star formation takes place; 2) In regions where $Q \gg 1$ cloud formation would be suppressed and only sporadic star formation, if any, should be present. This could occur in the outer regions of disks where the gas densities are low, or in the inner regions of galaxies where the threshold density is high; 3) In regions where $Q \sim 1$ the SFR will abruptly change from the unstable to stable regime. The slope of the star formation law should be very nonlinear, and cloud formation will be sensitive to local perturbations.

Over time the star formation acts to deplete the gas density until it approaches the $Q \sim 1$ stability limit. Hence a prediction of the model is that the radial distribution of gas surface density in most spiral disks should approximate a $1/r$ distribution near the stability limit. This in fact is observed (Figure 7), and hence the gravitational model provides a natural explanation for not only the distribution of star formation but also the distribution of gas in galaxies (Quirk 1972; Zasov & Simikov 1988). It also means that over most of the disk the gas lies in a quasi-stable regime, close to the transition between a roughly linear and highly linear Schmidt law.

3.3. OBSERVATIONAL TESTS OF THE Q MODEL

The Q model can be tested by comparing the observed threshold radii and surface densities in nearby galaxies with the values predicted from application of the Toomre criterion above. Such a comparison, for the galaxies shown in Figure 7, is shown in Figure 8 (Kennicutt 1989). Given the simple assumptions that went into the model (pure thin gas disk model, Toomre criterion, isothermal disk) the agreement is surprisingly good. We are currently repeating this analysis with a complete H I and CO-selected

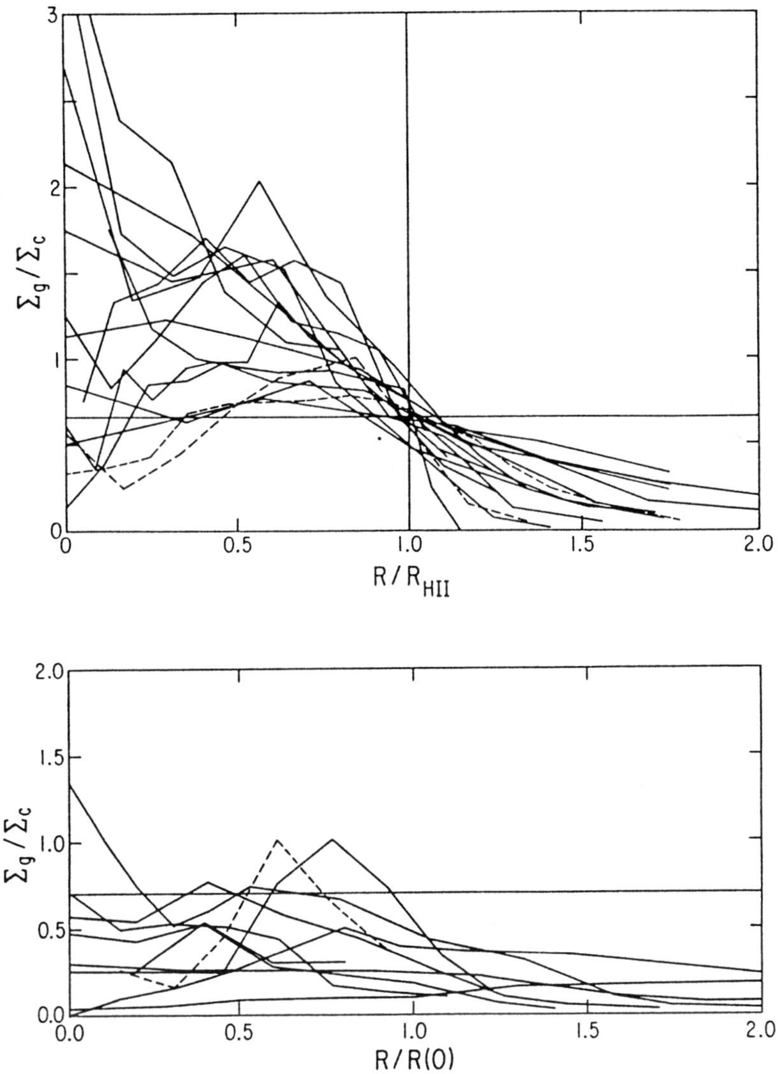

Figure 7. Variation of disk stability criterion $1/Q$ with radius for active star forming spirals (top) and low-SFR spirals (bottom). From Kennicutt (1989)

sample and better Hα data (Kennicutt & Martin 1997). In the meantime several investigators have tested the Q model with other types of galaxies.

Low surface brightness spirals (LSBs) and gas-rich S0 galaxies, with huge gas masses but negligible SFRs, are among the most enigmatic galaxies in the conventional Schmidt law picture. For example the prototype LSB,

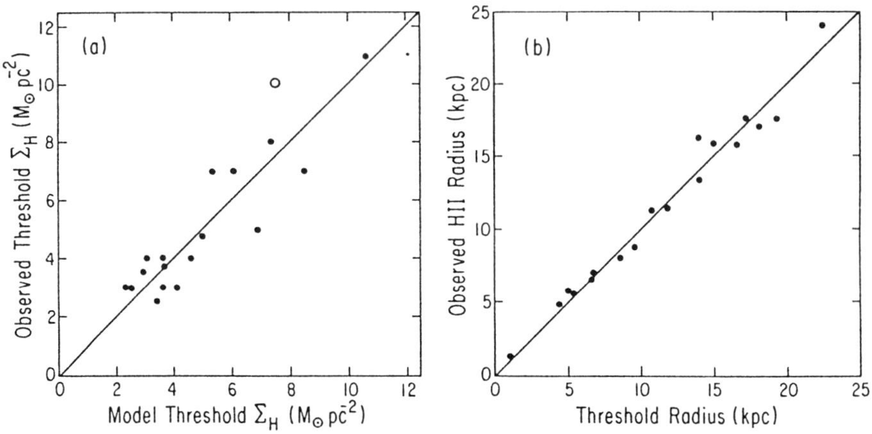

Figure 8. Comparison of predicted and observed threshold densities (left) and radii (right) for galaxies in the top panel of Fig. 7.

Malin 1, possesses a 200 kpc H I disk with a mass of $\sim 2 \times 10^{11} M_\odot$ ($H_0 = 75$), but with little detectable star formation (Impey & Bothun 1989). In the threshold picture the low SFRs are readily explained if the gas disk is sufficiently extended that $Q > 1.5$ everywhere. The nearly face-on orientation of Malin 1 makes it difficult to derive an accurate rotation curve (and hence Σ_c), but Impey & Bothun (1989) showed that its typical disk surface density is subcritical according to the Q criterion used here. Similar analyses of gas-rich S0 galaxies by Kennicutt (1989) and Eder (1990) reached the same conclusion.

The most comprehensive analysis of LSBs has been published recently by van der Hulst *et al.* (1993). These authors obtained VLA H I maps for 6 LSBs, and compared the radial variation in gas density with $\Sigma_c(r)$, following Kennicutt (1989). The resulting comparison is reproduced in Figure 9. In most of the galaxies $\Sigma_{HI}(r)$ indeed lies just below the $Q \sim 1.5$ threshold. In a few instances the H I density meets or exceeds the threshold value over a limited radial range, and in these cases the regions coincide with locations of H II regions. This conclusion could be modified if the galaxies contain substantial masses in molecular gas, but the absence of detected CO emission in most LSBs makes this unlikely. New observations of H I and CO in LSB galaxies (de Blok *et al.* 1996, de Blok and van der Hulst 1997) confirm this picture.

Taylor *et al.* (1994) have used VLA H I maps and optical imaging of 5 H II galaxies to compare the spatial extents of the star formation with that expected from a threshold model. The rotation curves in these low-mass irregular and spiral galaxies generally show solid-body rotation over most

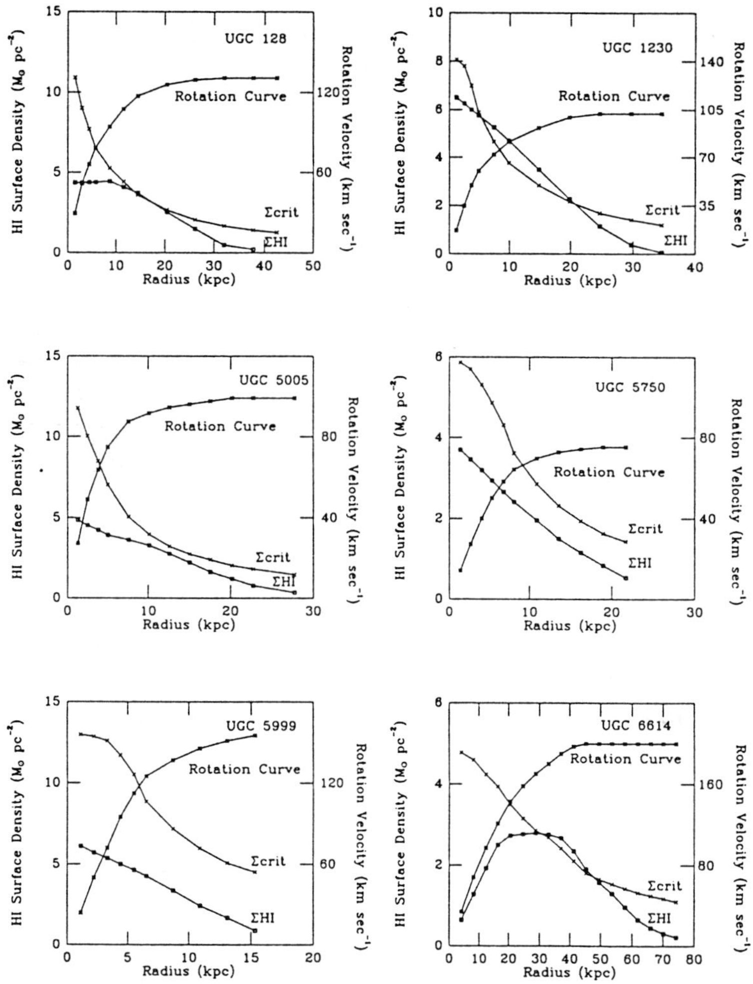

Figure 9. Rotation curves, H I density distributions, and critical density distributions for 6 low surface brightness galaxies, from van der Hulst *et al.* (1993).

of the optical disk, which yields a constant value of the critical density if the Toomre criterion is applied. However the galaxies show large spatial variations in H I density, so one can test for the presence of a threshold effect. Thresholds are seen in all cases, at densities roughly corresponding to those predicted by the Q model. However the range in critical densities cluster around a common value near 10^{21} H cm^{-2}, so other threshold mechanisms cannot be ruled out. Similar values for the threshold column densities were observed in a sample of nearby dwarf irregular galaxies by Skillman (1987), and Chamandaris *et al.* (1993) identified a threshold in the ring galaxy Arp 10. Kennicutt *et al.* (1995) have used Hα CCD images and H I maps of the Magellanic Clouds to perform the same test. They identify threshold densities of $\sim$0.7 and 2.0 $\times 10^{20}$ H cm^{21} for the LMC and SMC, respectively, compared to predicted values of $\sim$1.0–1.3 $\times 10^{21}$. It is not clear whether this disagreement indicates a breakdown in the Q model for these galaxies or is due to different gas velocity dispersions or some other factor. Overall it appears that the star formation properties of the irregular galaxies are in qualitative accord with the model, but the level of quantitative agreement may not be as good as for more massive spirals.

An especially powerful test for dynamically-induced threshold effects would be to identify galaxies with *azimuthal* variations in column density which cross the $Q = 1.5$ threshold, and test whether the SFR varies nonlinearly across this transition. Grand-design spirals offer an especially attractive laboratory for such tests, but until recently CO and H I data with sufficient spatial resolution were not available. Rand (1993) has used interferometer CO and H I maps of M51 to perform this test. He measured the variation in gas density and SFR (via Hα) across the spiral arms in M51, and compared the gas densities in the arm and interarm regions with the threshold densities expected from the Q criterion, taking into account the effects of the spiral density wave on the epicyclic frequency κ. He found that the gas surface densities in the arm and interarm regions bracketed the expected threshold value from the Q criterion, which is consistent with the strong arm/interarm contrast in Hα relative to CO. Uncertainties in several parameters, not the least the CO–H_2 conversion factor, may influence these conclusions, but Rand's analysis demonstrates the value of this approach, and it should be extended to other galaxies.

Despite these successes, the Q model is flatly contradicted by one class of galaxies— low-mass spirals. Two galaxies of this type, M33 and NGC 2403, are indicated by the dashed lines in Figure 7. Both exhibit subcritical surface densities over much of their inner disks, even though vigorous star formation is taking place in these regions. Another nearby spiral, NGC 300, appears to show similar behavior (Ferguson, unpublished). These spirals differ from their more massive counterparts in two significant respects; they have lower disk metal abundances, which might affect the C0–H_2 conversion ratio, and they exhibit slowly rising rotation curves over most of the inner disks. The former effect might lead to an underestimate of the gas densities in the inner disks, and explain the apparently subcritical densities in the active star forming regions. The latter effect might lead to a breakdown of the Toomre criterion for predicting the critical densities for cloud formation and star formation, as discussed below.

Another approach to testing the consequences of the star formation thresholds is to explore their effects on the star formation histories and chemical evolution of disks. An early exploration of this kind was made by Lacy & Fall (1985), who modelled the chemical evolution of the Galactic disk using a Schmidt law with a Q threshold. The most extensive analysis of this kind has been published by Wang & Silk (1994), who model the temporal evolution of the SFR and chemical enrichment in the Galactic disk using a complete star formation model which incorporates a Q threshold and self-regulation. Their model is successful at reproducing the observed SFR, the G-dwarf metallicity distribution, the local age-metallicity relation, and radial variations in gas density, SFR, and abundances. Chamcham *et al.* (1993) have constructed a similar model, but argue that gas flows and/or infall are needed to reproduce the abundance properties of the Galactic disk, an old problem with the conventional Schmidt law models. Firmani *et al.* (1996) follow the viscous evolution of disks incorporating a Q star formation parameter, and are able to produce exponential disks, flat rotation curves, and realistic Schmidt star formation laws. In different kind of application, the evolution of a starburst triggered by a high Q threshold has been discussed by Firmani & Tutukov (1993) and Elmegreen (1994a).

3.4. PHYSICAL NATURE OF THE THRESHOLDS

Although we refer to the observed cutoffs in star formation as "star formation thresholds," the underlying physical mechanism almost certainly involves regulation of some other process leading to star formation, such as the formation of gravitationally bound cloud complexes, the formation of molecular clouds, density wave triggering of cloud growth, or some other process. In this section we briefly discuss possible threshold mechanisms and alternatives which have been proposed to the Toomre Q criterion.

3.4.1. *Shear and tidal instabilities*

The physical motivation for applying the Safranov/Toomre criterion in Kennicutt (1989) was the idea that the onset of large-scale star formation might be dictated primarily by the growth of gravitational perturbations leading to cloud formation. However the Toomre criterion— which governs long-wavelength disk instabilities— may not describe these cloud growth instabilities in all conditions. Elmegreen (1993b) proposed an alternative critical density condition in which the epicyclic frequency κ is replaced with the shear rate, which in terms of the Oort constants is 2.5 A (*vs* $\kappa^2 = [4B(B-A)]$). Kenney *et al.* (1993) proposed a tidal criterion, in which κ^2 is replaced by $[3A(A-B)]$. Both of these critical densities converge to the Toomre value in the limit of constant rotation velocity (with a small scaling factor), but tend toward zero in the limit of solid body rotation (*vs* a constant critical density for the Q criterion). Hence the Elmegreen and Kenney *et al.* criteria predict much lower thresholds in the central parts of galaxies, or in low-mass spiral and irregular galaxies where there is relatively little shear. This could explain the high SFRs in spirals like M33, NGC 2403, or NGC 300, where there is an apparent failure of the Q model. However it does not explain the observation of star formation thresholds in Magellanic irregular galaxies (Skillman 1987, Taylor *et al.* 1994), which are accounted for quite readily by the Q model. Perhaps the best way to distinguish between these respective models is to measure the SFRs and gas densities near the centers of massive spiral galaxies, where the Q model predicts a much higher threshold density in most cases.

3.4.2. *Two-fluid and combined instabilities*

The simple Toomre criterion applied by Kennicutt (1989) and others applies strictly to a purely gaseous disk, and it is well known that the interaction between the stellar and gaseous disks can act to destabilize the gas disk (Jog & Solomon 1984). The consequences of adopting a more realistic two-fluid stability condition have been explored by Jog (1992), Elmegreen (1994b, 1995), Gammie (1996), and Fuchs (1996). While the threshold conditions predicted by these models are in qualitative accord with the simpler Toomre condition, Elmegreen (1994b) concludes that the differences in predicted threshold radii and densities should be measurable. The paucity of data on the velocity dispersions of the stellar *and* gas disks of nearby galaxies hampers such a test at present. Elmegreen (1991) and Gammie (1996) also consider a wide range of combined instability mechanisms for galactic gas layers, including gravitational, shear, magnetic Rayleigh-Taylor, and thermal instabilities. The interplay between these different instabilities may regulate star formation so that $Q \sim 1$ (Elmegreen 1991).

3.4.3. *Spiral density wave thresholds*

If spiral density waves are the primary triggering agent for large-scale star formation, they could produce threshold features as well (Wyse 1986, Wyse & Silk 1987). These authors have proposed a modified Schmidt law of form:

$$R = A\Sigma^n(\Omega - \Omega_p)$$

where Ω and Ω_p are the disk and spiral pattern angular frequencies. Such a model will produce a cutoff in star formation at the corotation radius, and a roughly $1/r$ term from $\Omega - \Omega_p$ inside this radius. Kennicutt (1989) has argued that the range of star formation properties in grand-design and flocculent spirals suggests a decoupling between the stellar disk instabilities (which lead to spiral density waves) and the gas instabilities (which lead to large-scale star formation), but this conjecture has not been tested observationally.

3.4.4. *Molecular vs. gravitational thresholds*

The close association of star formation with molecular cloud complexes implies that the formation of gravitationally bound clouds in itself is not a sufficient condition for large-scale star formation, and it raises the question of whether it is the formation of molecular gas, rather than the formation of gravitationally bound clouds, that is the crucial regulator of star formation. Are the formation of bound clouds and molecular clouds manifestations of the same threshold process, or are gravitational and molecular formation thresholds distinct processes which are decoupled from each other (and from star formation) in some environments?

An insightful exploration of the interplay between the formation of gravitational bound clouds and molecular clouds has been published by Elmegreen (1993a). In the solar neighborhood the formation of gravitationally bound clouds and molecular clouds are closely associated, because the threshold column densities for forming both objects are nearly the same. However the dependence of these thresholds on the interstellar pressure and metallicity are quite different, so in other environments the formation of bound clouds and molecular gas may be decoupled.

These ideas have interesting implications for the interpretation of the observed star formation thresholds in different types of galaxies. For example in disks with conditions similar to those found in the solar neighborhood, the critical densities for bound cloud formation and molecule formation are similar, and hence the observed star formation thresholds probably coincide with the threshold which separates gravitationally bound *and* molecular clouds from diffuse (pressure-supported) atomic gas. However these conditions do not hold in low-mass irregular galaxies, where low ISM pressures and low metallicities lead to threshold densities for molecular cloud formation which are much higher than the corresponding threshold for the formation of gravitationally bound clouds. In those environments gravitational instabilities will produce bound *atomic* cloud complexes, which later form molecular cores and stars within the larger atomic clouds (Elmegreen 1993a). At the other extreme, in metal-rich and/or high pressure interstellar environments, such as that found in the inner Galaxy or in massive gas-rich spirals such as M51, conditions will favor the formation of molecular gas at column densities which are too low to form gravitationally bound clouds. In these galaxies we would expect the observed star formation thresholds to occur in a largely molecular medium, consisting of bound molecular clouds within a substrate of diffuse molecular and atomic gas. In such galaxies it is the gravitational threshold, not the molecular formation (self-shielding) threshold, which is critical for initiating large-scale star formation. However in low-metallicity or low-pressure environments, such as in dwarf galaxies or the outer disks of spirals, gravitational stabilities may be secondary to molecule formation in leading to efficient large-scale star formation (*cf.* Lin & Murray 1992).

There is some observational evidence to support this basic picture. CO observations of the inner Galaxy reveal a diffuse molecular medium surrounding the molecular clouds (e.g., Polk *et al.* 1988, Stark *et al.* 1991), and the center of the nuclear starburst galaxy NGC 1808 shows evidence for a similar medium (Aalto *et al.* 1994). Observations of the CO and cold H I in the Magellanic Clouds (e.g., Israel *et al.* 1993; Dickey *et al.* 1994) are consistent with the cloud structure described by Elmegreen (1993a), though the interpretation of these data is complicated by possible changes in the CO/H_2 coupling in these metal-poor environments. It should be possible to separate these gravitational and molecular thresholds by obtaining sensitive maps of the CO, H I , and Hα in nearby galaxies which span a broad range of interstellar pressures, metal abundances, and rotation curves.

4. Regulation of the SFR

At high gas densities most galaxies show remarkably similar star formation laws, roughly following a Schmidt law with index $N \sim 1.3$–1.5, and constant zeropoint (see Figure 6 and Kennicutt 1989). This strongly hints toward some sort of regulation of the SFR, either from the gas–star conversion process itself, or from a self-regulating interaction of the star formation and the ISM. In this section I discuss some of the current ideas on the underlying physics of the star formation law in this high-density regime.

4.1. ISM-STAR INTERACTIONS

If one adopts as a working hypothesis that the SFR is determined purely by the efficiency and time scale for converting gas to stars, then one can identify several relevant processes that might produce a Schmidt law similar to what is observed. A thorough discussion of this subject is given by Larson (1987). For example, a linear Schmidt law can be obtained simply by postulating that the SFR is regulated by a constant star formation efficiency. Combining the local efficiency of $\sim$5% in interstellar clouds with a typical dynamical time scale of 10^8 years for disks yields a mean SFR per unit gas density which is very close to the mean global value observed in nearby spirals (e.g., Figure 2). A more nonlinear Schmidt law (e.g., $N \sim 2$) would result if the time scale for forming clouds τ in an ISM with surface density Σ scales with Σ^{-1}. Then the SFR per unit area will scale as $\Sigma_{\rm gas}/\tau \propto \Sigma_{\rm gas}^2$ (Larson 1987). Unfortunately the scale-free nature of the Schmidt power law makes it difficult to identify observationally a single underlying physical mechanism. However as discussed by Larson (1987), there are several processes, including gravitational instabilities, density-wave compression, accretion in a shear flow, and cloud coagulation, which have roughly the time scales and density dependence needed to explain qualitatively the form of the observed Schmidt law.

4.2. STAR-ISM INTERACTIONS AND FEEDBACK

An alternative possibility is that the SFR above the threshold is not determined uniquely by the gas density, but rather by a self-limiting feedback process between the ISM and the SFR itself. Self-regulated star formation models have been popular for some time. Proposed regulation mechanisms have included disk Ly-α opacity (Cox 1983) and energy injection from stellar winds and/or supernovae (e.g., Franco & Cox 1983; Dopita 1985, 1990; Parravano 1989; Burkert *et al.* 1992; Firmani & Tutukov 1992, 1994; Wang & Silk 1994; Dopita & Ryder 1994; Köppen *et al.* 1995). The combined effects of self-regulation and Q thresholds have been explored by Wang & Silk (1994) and Firmani & Tutukov (1994). A thorough review of the subject has been published recently by Shore & Ferrini (1994), and I will only highlight a few examples that are especially relevant to interpreting the large-scale star formation law.

Although the details of the self-regulation mechanisms vary between different models, in most cases the limiting SFR is determined by the balance between the turbulent ISM pressure or heating introduced from stellar winds and supernovae and the ambient thermal or binding pressure of the warm ISM. If the input from star formation is much below the equilibrium pressure the gas disk contracts, promoting cloud growth and a higher SFR. If the SFR exceeds its equilibrium value the excess energy input will disperse clouds or even the disk itself, cutting off star formation until an equilibrium condition is achieved again. This process often leads to a Schmidt-like SFR–density law, with N typically ranging from 1.3–2 (e.g., Parravano 1989; Dopita 1990; Köppen *et al.* 1995; Friedli & Benz 1995). Because the binding pressure of the ISM also scales with the stellar surface density, some models predict a compound Schmidt law of the form:

$$\Sigma_{sf} \propto \Sigma_{\rm gas}^N \, \Sigma_*^M$$

where $\Sigma_{\rm gas}$ and Σ_* are the gas and stellar surface densities, respectively (e.g., Dopita 1985, 1990).

Many of these models contain too many physical variables and free parameters to offer a robust observational test, but some predict an explicit form for the star formation law which can be compared to observations. For example, the models of Dopita (1985, 1990) predict compound Schmidt laws of the form given above, with $N = 1$–1.33 and $M \sim 1$. Dopita & Ryder (1994) also consider a law of the form $\Sigma_{sf} \propto \Sigma_{\rm gas}\Omega$, where Ω is the angular rotation speed.

Observational evidence for these types of law is presented by Ryder (1993), who finds that the ratio of the SFR to H I surface density is correlated with local *stellar* surface density. When the latter is parametrized in terms of I-band surface brightness they find a best fit to a law:

$$SFR \propto \Sigma_{\ddot{\rm I}0.9}\ \Sigma_I^{0.64}$$

similar to that predicted by their self-regulating models. Kennicutt & Martin (1997) have independently tested for a compound Schmidt law for the galaxy sample shown in Figure 2, by correlating the mean Hα surface brightness with the product of the mean gas density (H I + H_2) and the stellar (V) surface brightness. We do find an excellent correlation with slope near unity, as predicted by the pressure-regulated model (e.g., Dopita 1985), but the fit is no better than a simple Schmidt law with $N > 1$. A problem we encountered with the interpretation of both of these results is that the stellar surface brightness in most disks is strongly correlated with the surface density of molecular gas— CO disks are exponential with roughly the same scale length as the stars— so it is difficult to determine *a priori* whether a correlation of SFR with stellar density is actually due to a dependence on Σ_* independent of $\Sigma_{\rm gas}$ or simply a dependence on $\Sigma_{\rm gas}$ to a higher power. This could be tested in principle by analyzing a sample of galaxies with similar stellar disk properties but a wide range of gas disks or *vice versa.* Unfortunately the range of stellar disk properties in most available samples is too limited for this purpose, and the tendency of gas disks to follow a $Q \sim 1$ profile forces a coupling between the stellar and gaseous disk profiles (Kennicutt 1989), which also limits the range of disk properties which can be readily observed.

Perhaps the most impressive demonstration of the importance of stellar wind and supernova feedback for the SFR and disk evolution comes from numerical simulations. When the formation of disk galaxies is simulated using particle/gas models which include the effects of self-gravity, gas cooling, and star formation, but without mechanical feedback from star formation, the resulting disks are too condensed, and star formation tends to proceed more rapidly than is observed (Navarro & Benz 1991). More recent models which incorporate explicitly the effects of stellar feedback show that the global SFR is a very sensitive to the mechanical energy injection rate (Navarro & White 1994; Friedli & Benz 1995). This is illustrated in Figure 10, which shows the dependence of the equilibrium SFR on the supernova rate and mechanical energy injection rates, for a set of disk

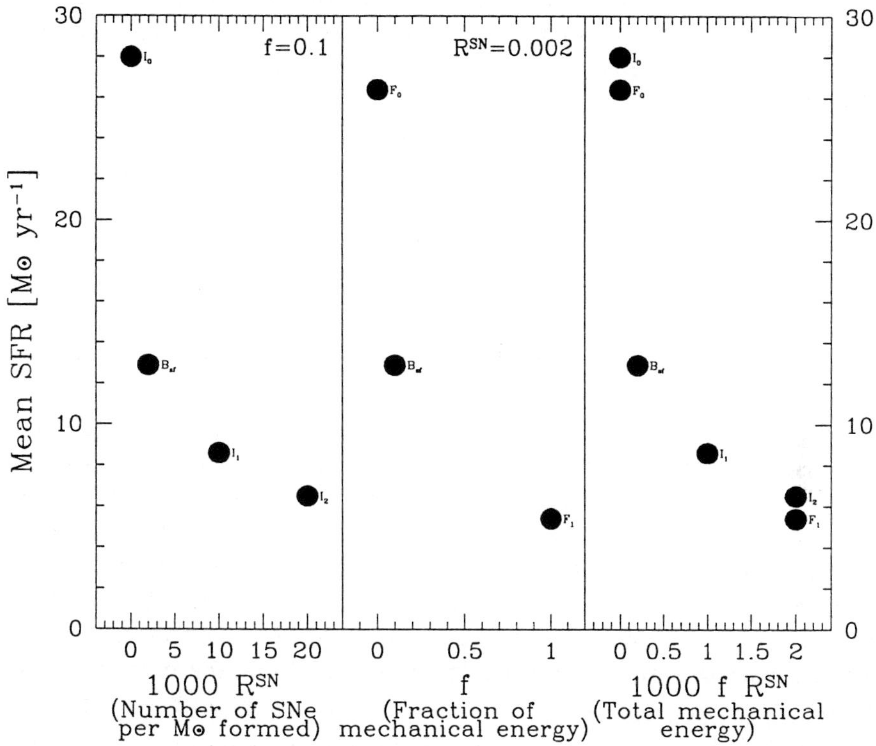

Figure 10. Dependence of the SFR per unit gas mass on the mechanical energy injection rate from massive stars and supernovae, from models of Friedli & Benz (1995).

simulations from from Friedli & Benz (1995). Similar results have been derived by Navarro & White (1994). The models also demonstrate that the feedback affects the spatial distribution of the star formation; for example the spatial concentration of the star formation in spiral arms can be wiped out if the feedback efficiency is too high. Many of the simulations bear a striking resemblance to real galaxies, and some are capable of reproducing the observed star formation law (Friedli & Benz 1995).

The critical parameter in these models is the ratio of *mechanical* energy imparted to the ISM from a generation of stars to the initial mass of stars formed, the "feedback efficiency" referred to above. The *total* energy input from stars and supernovae can be calculated in principle using stellar models (e.g., Leitherer & Heckman 1995), but the results are sensitive to metallicity and the temporal properties of the star formation event, and are sensitive to uncertainties in the stellar evolution models themselves. A further uncertainty is the fraction of the stellar wind and supernova energy which is transferred into bulk kinetic energy in the ISM (thermal energy imparted to the ISM is quickly radiated away and does not influence the SFR). This quantity is not easily calculated and may well depend on the clustering properties of the star formation (e.g., Mac Low & McCray 1989; Norman & Ikeuchi 1989; Heiles 1990). Observations of the kinematics of the ISM in star forming regions may provide more direct measurements of these energy injection parameters and their variation in different environments (e.g., Chu & Kennicutt 1994). Despite these uncertainties the simulations clearly demonstrate the importance of the feedback from star formation on the formation, evolution, and structure of disk galaxies.

As impressive as these state-of-the-art 2D simulations are, computational limitations restrict the range of parameter space which can be explored. In particular, relatively simple "recipes" must be adopted for the SFR as a function of gas density, the growth rates of clouds and star formation, and the transfer of energy from stars back to the ISM. The effects of varying these underlying assumptions have been explored in a series of papers by Struck-Marcell & Scalo (1987) and references therein. Their single-zone "Oort models" lack the sophistication of the modern 2D simulations, but this approach has allowed them to adopt more realistic algorithms for cloud formation and destruction, star formation, and the nonlinear growth of coherent structures and starbursts from delayed feedback effects.

One of the most significant results of this work was the discovery that the nature of the star–ISM feedback and the resulting temporal evolution of the system depends critically on the time delays between cloud formation, star formation, and the subsequent disruption of the surrounding ISM by winds and supernovae. For instantanous models (i.e., short delay time for feedback) the systems exhibit negative feedback, similar to the behavior exhibited in the 2D models described above (e.g., Fig. 10); local perturbations to the SFR tend to be damped out. However as the delay time between cloud formation and stellar energy injection becomes comparable to the growth time for forming clouds, the systems become unstable and the models display positive feedback, first exhibiting a limit cycle behavior, then a chaotic series of bursts punctuated by quiescent periods as the feedback delays are increased further. Bursts can also be introduced in other ways, for example by introducing gas into the system as a function of time (Vázquez & Scalo 1989). Hence the negative feedback behavior seen in the 2D models occurs over only a limited set of conditions, and for many realistic parameters the qualitative character of the star-ISM interaction is radically different, with much of the star formation occuring in massive bursts. It is possible that some of the limit-cycle behavior exhibited in these models is an artifact of the single-zone boundary conditions (Scalo, private communication), but the qualitative results provide a reminder that more sophisticated 2D multi-particle models need to be performed over a wider range of input conditions and "recipes."

Taken together these simulations suggest that the effects of star formation on the ISM may be as important as the SFR–gas density law in driving the global star formation and evolution of galaxies. At present this subject is being driven by theory and simulations, in much the same way that the entire subject of star formation in galaxies was being driven by theory 20 years ago. Although the inherent complexity of star–ISM interactions on galactic scales is a formidable obstacle to observers, the potential for new physical insights into the regulation of the SFR and galaxy evolution make this an exciting and potentially rewarding field for future investigation.

This work was supported in part by the National Science Foundation through grant AST-9019150. I would also like to acknowledge my collaborators and graduate students Fabio Bresolin, Crystal Martin, and Sally Oey for their contributions to some of the results presented here.

References

Aalto, S., Booth, R. S., Black, J. H., Koribalski, B., and Wielebinski, R. 1994, *Ast.Ap.* , 286, 365.
Berkhuijsen, E. M. 1977, *Ast.Ap.* , 57, 9.
Boselli, A. 1994, *Ast.Ap.* , 292, 1.
Buat, V., Deharveng, J. M., and Donas, J. 1989, *Ast.Ap.* , 223, 42.
Buat, V. 1992, *Ast.Ap.* , 264, 444.
Burkert, A., Truran, J. W., and Hensler, G. 1992, *Ap.J.* , 391, 651.
Chamcham, K., Pitts, E., and Tayler, R.J. 1993, *M.N.R.A.S.* , 263, 967.
Charmandaris, V., Appleton, P. N., and Marston, A. P. 1993, *Ap.J.* , 414, 154.
Chu, Y.-H., and Kennicutt, R. C. 1994, *Ap.J.* , 425, 720.
Cowie, L. L. 1981, *Ap.J.* , 245, 66.
Cox, D. P. 1983, *Ap.J.* , 265, L61.
Davies, R. D., Elliott, K. H., and Meaburn, J. 1976, *Mem.R.A.S.* , 81, 89.
de Blok, W.J.G., McGaugh, S.S. and van der Hulst J.M. 1996, *M.N.R.A.S.* , 283, 18.
de Blok, W.J.G. and van der Hulst J.M. 1997, *M.N.R.A.S.* , submitted.
Deharveng, J. M., Sasseen, T. P., Buat, V., Bowyer, S., Lampton, M., and Wu, X. 1994, *Ast.Ap.* , 289, 715.
Dickey, J. M., Mebold, U., Marx, M., Amy, S., Haynes, R. F., and Wilson, W. 1994, *Ast.Ap.* , 289, 357.
Dopita, M. A. 1985, *Ap.J.* , 295, L5.
Dopita, M. A. 1990, in *The Interstellar Medium in Galaxies*, eds. H. A. Thronson and J. M. Shull, Kluwer, p. 437.
Dopita, M. A., and Ryder, S. D. 1994, *Ap.J.* , 430, 163.
Eder, J. A. 1990, PhD thesis, Yale University.
Elmegreen, B. G. 1979, *Ap.J.* , 231, 372.
Elmegreen, B. G. 1991, *Ap.J.* , 378, 139.
Elmegreen, B. G. 1993a, *Ap.J.* , 411, 170.
Elmegreen, B. G. 1993b, Proc. 4th EIPC Conference, p. 335.
Elmegreen, B. G. 1994a, in *Violent Star Formation*, ed. G. Tenorio-Tagle, Cambridge Univ Press, p. 220.
Elmegreen, B. G. 1994b, *Ap.J.* , 427, 384.
Elmegreen, B. G. 1995, *M.N.R.A.S.* , 275, 944.
Firmani, C., and Tutukov, A. V. 1992, *Ast.Ap.* , 264, 37.

Firmani, C., and Tutukov, A. V. 1993, *Rev. Mex. Astr. Ap.*, 27, 129.

Firmani, C., and Tutukov, A. V. 1994, *Ast.Ap.* , 288, 713.

Firmani, C., Hernandez, X., and Gallagher, J. 1996, *Ast.Ap.* , 308, 403.

Franco, J., and Cox, D. P. 1983, *Ap.J.* , 273, 243.

Freedman, W. L. 1984, PhD thesis, University of Toronto.

Friedli, D., and Benz, W. 1995, *Ast.Ap.* , 301, 649.

Fuchs, B. 1996, *M.N.R.A.S.* . 278, 985.

Gammie, C. F. 1996, *Ap.J.* , 462, 752.

Guiderdoni, B. 1987, *Ast.Ap.* , 172, 27.

Heiles, C. 1990, *Ap.J.* , 354, 483.

Hunter, D. A., and Gallagher, J. S. 1986, *P.A.S.P.* , 98, 5.

Impey, C., and Bothun, G. 1989, *Ap.J.* , 341, 89.

Israel, F. P. *et al.* 1993, *Ast.Ap.* , 276, 25.

Jog, C. J., and Solomon, P. M. 1984, *Ap.J.* , 276, 127.

Jog, C. 1992, *Ap.J.* , 390, 378.

Kenney, J. D. P., Carlstrom, J. E., and Young, J. S. 1993, *Ap.J.* , 418, 687.

Kennicutt, R. C. 1983, *Ap.J.* , 272, 54.

Kennicutt, R. C. 1989, *Ap.J.* , 344, 685.

Kennicutt, R. C. 1990, in *The Interstellar Medium in Galaxies*, eds. H. A. Thronson and J. M. Shull, Kluwer, p. 405.

Kennicutt, R. C., Tamblyn, P., and Congdon, C. W. 1994, *Ap.J.* , 435, 22.

Kennicutt, R. C., Bresolin, F., Bomans, D. J., Bothun, G. D., and Thompson, I. B. 1995, *A.J.* , 109, 594.

Kennicutt, R. C., and Martin, C. L. 1997, in preparation.

Knapen, J.H. and Beckman, J.E. 1996, *M.N.R.A.S.* , 283, 251.

Köppen, J., Theis, Ch., and Hensler, G. 1995, *Ast.Ap.* , 296, 99.

Lacey, C. G., and Fall, S. M. 1985, *Ap.J.* , 290, 154.

Larson, R. B. 1983, *Highlights in Astr.*, 6, 191.

Larson, R. B. 1987, in *Starbursts and Galaxy Evolution*, eds T. X. Thuan, T. Montmerle, and J. Tran Tranh Van, Editions Frontierières, p. 467.

Larson, R. B. 1988, in *Galactic and Extragalactic Star Formation*, eds. R. E. Pudritz and M. Fich, Kluwer, p. 435.

Leitherer, C., and Heckman, T. M. 1995, *Ap.J.Suppl.* , 96, 9.

Lin, D. N. C., and Murray, S. D. 1992, *Ap.J.* , 394, 523.

Mac Low, M.-M., and McCray, R. 1988, *Ap.J.* , 324, 776.

Madore, B. F. 1977, *M.N.R.A.S.* , 178, 1.

Navarro, J. F., and Benz, W. 1991, *Ap.J.* , 380, 320.

Navarro, J. F., and White, S. D. M. 1994, *M.N.R.A.S.* , 265, 271.

Norman, C. A., and Ikeuchi, S. 1989, *Ap.J.* , 345, 372.

Quirk, W. J. 1972, *Ap.J.* , 176, L9.

Parravano, A. A. 1989, *Ap.J.* , 347, 812.

Polk, K. S., Knapp, G. R., Stark, A. A., and Wilson, R. W. 1988, *Ap.J.* , 332, 432.

Rand, R.J. 1993, *Ap.J.* , 410, 68.

Rogstad, D. H., Lockhart, I. A., and Wright, M. C. H. 1974, *Ap.J.* , 193, 309.

Ryder, S. D. 1993, PhD thesis, Australian National University

Safronov, V. S. 1960, *Annales d'Astrophys*, 23, 979.

Schmidt, M. 1959, *Ap.J.* , 129, 243.

Shaya, E. J., and Federman, S. R. 1987, *Ap.J.* , 319, 76.

Shore, S. N., and Ferrini, F. 1994, *Fund. Cosm. Phys.*, in press

Silk, J. 1987, in *Star Forming Regions*, eds. M. Peimbert and J. Jugaku, Reidel, p. 663.

Skillman, E. D. 1987, in *Star Formation in Galaxies*, ed. C. J. Lonsdale Persson, NASA Conf. Publ. 2466, p. 263.

Stark, A. A., Gerhard, O. E., Binney, J., and Bally, J. 1991, *M.N.R.A.S.* , 248, 14P.

Struck-Marcell, C., and Scalo, J. M. 1987, *Ap.J.Suppl.* , 64, 39.

Taylor, C. L., Brinks, E., Pogge, R. W., and Skillman, E. D. 1994, *A.J.* , 107, 971.

Tilanus, R. P. J., and Allen, R. J. 1989, *Ap.J.* , 339, L57.
Toomre, A. 1964, *Ap.J.* , 139, 1217.
Tosa, M., and Hamajima, K. 1975, *P.A.S.J.* , 27, 501.
van der Hulst, J. M., Skillman, E. D., Kennicutt, R. C., and Bothun, G. D. 1987, *Ast.Ap.* , 177, 63.
van der Hulst, J. M., Skillman, E. D., Smith, T. R., Bothun, G. D., McGaugh, S. S., and de Blok, W. J. G. 1993, *A.J.* , 106, 548.
van der Kruit, P. C., and Shostak, G. S. 1984, *Ast.Ap.* , 134, 258.
Vázquez, E. C., and Scalo, J. M. 1989, *Ap.J.* , 343, 644.
Wang, B., and Silk, J. 1994, *Ap.J.* , 427, 759.
Wyse, R. F. G. 1986, *Ap.J.* , 311, L41.
Wyse, R. F. G., and Silk, J. 1987, *Ap.J.* , 339, 700.
Zasov, A., and Simikov, S. 1988, *Astrophysica*, 29, 190.

SUBJECT INDEX

OBJECT INDEX

LIST OF SYMBOLS

A_V extinction in V
A Oort constant A
B Oort constant B
ΔV line profile width
D_{25} de Vaucouleurs diameter
$E(\lambda - V)/E(B - V)$ color excess
f filling factor
G_{surf} H_2 photodissociation rate
G gravitational constant
h_e scale height
$I_{CO} = \int T_b dv$ integrated CO brightness
k Bolzmann's constant
L_X X-ray luminosity
L_{50} ionizing luminosity in units of 10^{50} photons s^{-1}
$L_{CO} = \pi R^2 T_b S$ CO luminosity
L_B blue optical luminosity
L_{FIR} far infrared luminosity
$L_{H\alpha}$ Hα luminosity
$\mathcal{L}_B$ blue optical luminosity
$\mathcal{M}_{HI}$ H I masss
$\mathcal{M}_T$ total mass
$\mathcal{M}_{dark}$ mass in dark matter
$\mathcal{M}_{lum}$ mass in luminous matter
$\mathcal{M}/\mathcal{L}$ mass to luminosity ratio
M_d dust mass
$M_{vir} \simeq R^2/G$ virial mass
$M(H_2)$ mass of H_2
$M_\odot$ solar mass
M_B absolute blue magniude
$M_T(R)$ total mass
m_H mass of the hydrogen atom
N_{HI} H I column density
n_{gas} mean gas density
N_{H_2} H_2 column density
N_H hydrogen column density
n gas density
p_{eff} effective yield
q ionization parameter
Q Toomre's stability parameter
R galactocentric radius
r galactocentric radius
R_H rate constant for H_2 formation
R_{25} de Vaucouleurs radius
$R_{25}^{b,i}$ inclination–corrected R_{25}
R_s Strömgren radius
r_c core radius
R_{23} Oxygen/H_β line ratio (ch. 6)
R_a effective aperture
R_d effective radius of the disk
R_{sp} effective radius of the spheroid
R_c corotation radius
S flux density
$\int SdV$ emission line flux
T_d dust temperature
T_c color temperature
T_{eq} equivalent black body temperature
T_b brightness temperature
T_g gas temperature
T_X X–ray temperature
T morphological type (ch. 6)
U Ionization Parameter
V velocity
V_c circular rotation velocity
X_{edge} the ionization fraction at the edge of the model
X CO–H_2 conversion factor
z distance perpendicular to the plane
z_0 vertical scale height
Z metallicity
$Z_\odot$ solar metallicity

κ optical depth (ch. 2)
κ epicyclic frequency (ch. 3, 7)
λ wavelength
μ mean molecular weight
ν wavelength
ρ density
ρ_{sf} volume density of young stars
ρ^n_{gas} volume density of gas
σ rms error
$\sigma_{\rm v}$ velocity dispersion
σ_z velocity dispersion in the z direction
$\Sigma_{\rm HI}$ H I surface density
$\Sigma_{\rm X}$ X-ray surface brightness
$\Sigma_{\rm gas}$ gas surface density
Σ_{sf} surface density of young stars
Σ_c critical surface density for star formation
Σ_* surface density of all stars
Σ_I surface brightness in I
τ time scale for forming molecular clouds
Ω angular rotation speed
Ω_p pattern speed